Praise for *The Secret Life of Pigs*

"This is not a book that talks only about the life of pigs in a sanctuary, it is a book that immerses us in a world where the ordinary becomes magical. A book of interspecies relationships, of funny anecdotes and some sad ones too, of transformations, of reencounter, of empathy, of courage, of overcoming obstacles, of rethinking everything we have learned, that which is so hard for us to banish from our kingdoms. A fundamental book to begin to build another humanity, a fairer one for all beings that inhabit this planet."—**Malena Blanco**, writer, animal rights activist and co-founder of Voicot

"This eye-opening book gives readers an in-depth look at the tireless work of those who dedicate their lives to our kindred animals. Richard Hoyle and Anita Krajnc send a clear and passionate message that every act of caring, whether through creating sanctuary, bearing witness, or participating in positive activism, raises the possibility for humans to truly envision a more peaceful world. Anita's retelling of her trial and the tragic killing of Regan Russell remind us that activists who bring the plight of animals to our attention often do so at great sacrifice and peril. This is an inspiring and impactful book."—**Dr. Joanne Kong**, editor of *Vegan Voices: Essays by Inspiring Changemakers*

"This book is a powerful testament to the majestic nature of animal sentience. Our relationship with these creatures is a reflection of who we are as people. These stories elicit emotion, compassion and give us the inspiration to be better people. The vegan message is a powerful tonic for our troubled times and our collective efforts for animal liberation is a vital key to the future of humanity and all life on earth."—**Robbie Lockie**, co-founder and director of Plant Based News

"What a brilliant book! Two amazing storytellers, each in their own style, showing how animals bring out the best of us—and how we can become better advocates for them. You will laugh, cry, think, and not let go until you reach the last page!"—**Dr. Camila Perussello**, author of *Food for Thought: Planetary Healing Begins on Our Plate*

"This is a riveting account of humans seeing pigs for who they really are: highly intelligent, social, emotional beings with the capacity to love, fear, and even forgive humans. It's also the fascinating tale of Anita Krajnc, who said 'enough' to society's systematic torture of pigs and got arrested and prosecuted for it, sparking a global movement."—**Jane Velez-Mitchell**, founder of UnchainedTV

The Secret Life of Pigs

Stories of Compassion and the Animal Save Movement

Richard Hoyle & Anita Krajnc

Lantern Publishing & Media • Woodstock and Brooklyn, NY

2022
Lantern Publishing & Media
PO Box 1350
Woodstock, NY 12498
www.lanternpm.org

Cover design by Emily Lavieri-Scull
Editing by Liza Barkova, Pauline Lafosse, and Hanh Nguyen

Printed in the United States of America

Library of Congress Cataloging-in-Publication Data

Names: Hoyle, Richard, 1948- author. | Krajnc, Anita, author.
Title: The secret life of pigs : stories of compassion and the Animal Save Movement / Richard Hoyle & Anita Krajnc.
Description: Woodstock : Lantern Publishing & Media, 2023. | Includes bibliographical references.
Identifiers: LCCN 2022026923 (print) | LCCN 2022026924 (ebook) | ISBN 9781590566664 (paperback) | ISBN 9781590566671 (epub)
Subjects: LCSH: Hoyle, Richard, 1948- | Swine as pets—Virginia—Anecdotes. | Swine as pets—Tennessee—Anecdotes. | Human-animal relationships—Anecdotes. | Pig Preserve (Pig sanctuary)—History. | Food animals—United States—Moral and ethical aspects. | Animal rescue—United States—Anecdotes. | Animal welfare—United States—Societies, etc. | BISAC: NATURE / Animal Rights | POLITICAL SCIENCE / Civil Rights
Classification: LCC SF395.6 .H69 2023 (print) | LCC SF395.6 (ebook) | DDC 636.4/0887—dc23/eng/20220629
LC record available at https://lccn.loc.gov/2022026923
LC ebook record available at https://lccn.loc.gov/2022026924

Contents

To the millions upon millions of pigs bred, born, and raised in unbelievably horrible conditions only to be brutally and inhumanely slaughtered without ever having seen sunlight, breathed fresh air, or felt green grass under their feet; to the billions who have died without ever having experienced a bit of human affection, kindness, or love; and to the billions whom we could not save . . . we dedicate this book, along with our heartfelt prayers and tears. We have not forgotten you. We are trying.

Acknowledgements

Richard Hoyle

AT THE TOP OF MY list of "things I will never do" in life, running a pig sanctuary had to be number one. Item number two on that list would have to be "writing a book about running a pig sanctuary."

Not too many years after we started our first sanctuary in Culpeper, VA, people began asking me if I was ever going to write a book about my experiences, including my ever-growing knowledge and understanding of pigs. I always laughingly deflected their questions, partly because I didn't think what I was doing was particularly noteworthy or unique, and partly because I wasn't sure whether rescuing and caring for pigs was a previously undiagnosed behavioral abnormality or a calling from on high. After thirty-six years, I'm still not sure of the answer to that question.

For several years, I wrote newspaper articles for our sanctuary's quarterly newsletter. I also wrote a recurring column entitled "Life at the Sanctuary" for *The Herd*, a modest little newsletter put out by a fellow pig sanctuary director in New Jersey named Marty Piatkowsky. The goal of both of these publications was to educate pet-pig guardians and the few interested people in the general public about the mission of pig sanctuaries and to try to pass on—mostly through the use of humor—some insight into what running a small, grassroots pig sanctuary was all about. Both of these writing gigs petered out after a couple of years and, to tell the truth, I thought no more about doing any writing until Anita Krajnc from the Animal Save Movement began harping on me in earnest. A word to the wise: never play poker with Anita. And if you do, don't, under any circumstances, try to bluff her. I used the excuse that there was no infrastructure available to me to write a book and, within a matter of a few months, she had arranged a sponsor, a ghost writer, and a publisher for the book. In short . . . she called my bluff.

There are a number of people and organizations that need to be recognized and thanked either for their help with the book or for their unflagging assistance over many years to me, to my sanctuary, and most importantly, to the pigs.

First, my sincere thanks to **Anita Krajnc**, the founder of the Animal Save Movement, who is always the smartest and most focused person in the room—any room. Anita single-handedly started a worldwide movement that has swept up millions of people in a multitude of countries to witness the suffering of and fight for the rights of pigs and all other farmed animals. If you look in any dictionary under "hero" or "heroine," you will probably find a photo of Anita.

Knowing that all of us who labor in the wonderful world of farmed-animal rescue tend to suffer from chronic pecuniary strangulation, Anita found an anonymous "sponsor" whom she convinced to help fund this book project, and without whom it would have certainly been dead in the water. I suspect that this donor probably got swept up by "Hurricane Anita" and her unremitting zeal for the project. And while I have no idea who this trusting and dedicated soul may be, I hope that we did you proud and that you feel, once the book is published, your generosity was well invested and that your trust in me was well placed.

I have never met **Brian Normoyle** or any of the numerous elves and gremlins who labor at Lantern Publishing and Media in Brooklyn, NY. I'm not even sure what all they do other than it's pretty much everything it takes (minus the actual writing) to turn a book from an idea into the finished product on the bookshelf. These wizards practice their literary alchemy and logistical sorcery behind the scenes, but their courage in agreeing to publish a book written by an unknown former Marine and retired firefighter/paramedic is either prophetic or really crazy . . . I guess time will tell. But thanks for taking a chance on me.

With Irish blood on both sides of my family, I have been blessed (cursed?) with the Gift of Blarney, which I once heard defined as "bullshit spread so thinly and so eloquently as to be imminently believable." And, like most true Irishmen, I have no idea how to make a long story short. My wife herself has long cautioned my friends never to ask me the time, as I would promptly tell them how to build a clock. The Herculean task of taking thirty-six years' worth of largely emotional and incoherent rambling on my part and molding, shaping, and kneading it all into a coherent, poignant, and well-organized book fell to **Sylvia Fraser**, my designated ghost writer. As the interviewing and writing process unfolded, it was obvious that Sylvia was both a true professional in her trade and a stern taskmaster. The finished product is, in no small measure, the result of Sylvia's skill, dedication, and patience with me over many months. Sylvia, you have my thanks, my respect, and my admiration.

Obviously, I have met and worked with a large number of people in my thirty-six–year career as a rescuer and sanctuary director. Were I to try to recognize and thank all—or even a significant number—of them, the acknowledgements would be longer than the body of the book. There are many friends, supporters, volunteers, interns, and professionals I will not be able to thank individually here . . . and for that I am sorry.

Early on, when Laura and I were trying to determine which road to take—to become a sanctuary or get out of the pig-rescue world—we stumbled upon two amazing and unlikely mentors. **Jim Brewer** and **Dale Riffle** of PIGS, A Sanctuary took us in tow for several years and shared their knowledge and their lives as the founders and directors of one of the first pig sanctuaries in the United States. Working closely with them, we discovered a lot of practices that we wanted to incorporate into our sanctuary and almost as many things that we wanted to do differently. Jim and Dale also started us on the path to a totally plant-based diet. These two dedicated men, along with their protégé **Susie Coston**, who later went on to work at Farm Sanctuary, gave us our first hint of the monastic lifestyle and level of dedication it took to run a grassroots or "mom-and-pop" sanctuary. We will be forever in their debt for their kindness and time they dedicated to helping us get our first sanctuary off the ground.

The success or failure of any sanctuary is always going to be contingent on the level of veterinary care provided to the animals. During my career, I have been blessed with three of the most amazing veterinarians:

Carole Nicholson, who ran a home veterinary practice out of her house in Stafford, VA, was our first vet. I would be hard-pressed to decide who knew less about the veterinary needs of pigs—us or Carole. And it took a great deal of cajoling and arm twisting for us to get her to take our fledgling herd on as clients. Veterinary procedures that we now accept as routine and mundane for pigs were new, unscripted, scary, and without precedent back then. There was virtually no body of veterinary medicine dealing with miniature pigs like the potbellied pigs. There were no resources to turn to for expert advice. There was no reliable Internet, huge search engines, or YouTube videos to help us research veterinary questions. We were totally "winging it" back in those days. To her credit, Carole was an excellent veterinarian who was also blessed with a great deal of common sense and the ability to say "I don't know" when it was appropriate. We lost a few pigs to ignorance and mistakes in the early years. But we saved many more than we lost and, more importantly, we began building a base of veterinary knowledge that would grow exponentially over the years. I owe an enormous debt of gratitude to Carole for her veterinary skills and her courage to do what no other vet would do.

As we began building Mini-Pigs, Inc. on our seventeen acres in Culpeper, VA, we also began searching for a vet in the area who would take on a rapidly growing pig sanctuary. **Dr. Tom Massie, Jr.** was a recent graduate of Virginia Tech School of Veterinary Medicine. He was running his first practice, Rosehill Veterinary Practice, LLC, out of his parents' home in Washington, VA. Tom and I developed an instantaneous rapport. Smart and aggressive, with the ability to think outside of the box, he served as our vet the entire time the sanctuary was located in Virginia, and we made a great team. I learned a tremendous amount from him, and between us, we tried some unique and cutting-edge veterinary techniques. Tom also instilled in me a new and deep respect for farm vets. Traveling in the "vet trucks" from farm to farm by himself, he dealt with the entire gamut of veterinary emergencies involving a wide range of animal species in the course of each day. His days often began before sunup and went well into the night as he answered routine and emergency calls. Ours was the vet–sanctuary director relationship that every sanctuary director dreams of but seldom finds. Tom challenged me every day to read up on different topics, research, ask questions, and come up with innovative ways to treat the pigs in my care. I cannot thank Tom enough for his services or his friendship. And I am glad to see that his practice has grown dramatically. I still consider him one of the finest vets and probably the most knowledgeable pig vet I have ever known. Thank you, Tom, for all you have done for our sanctuary and the pigs over the years.

Establishing The Pig Preserve in a rural area in the Tennessee mountains meant that our access to vets would be limited. But fate smiled on us when they put **Dr. Cindy Johnson** of Dogwood Animal Hospital in our path. This short dynamo of a woman is scary smart and a wonderfully resourceful vet who runs a both small- and large-animal practice in Jamestown, TN. While she is the only vet in her super-busy practice, she has a staff that has always been a joy to work with. Working with Cindy, we designed veterinary protocols for a host of porcine medical conditions—protocols that simply did not exist elsewhere. With the help of some local "human doctors," we developed treatments for various cardiac conditions in pigs using human as well as veterinary drugs. The fantastic health of our herd of over 175 pigs, in particular the low death rate and increased longevity of our farmed pigs, is directly attributable to Cindy's skill and expertise as a caring veterinarian. There are not enough thanks in the world to heap on Cindy and her staff at Dogwood Animal Hospital.

Dr. Sarel van Amstel, a professor emeritus of large-animal care who taught for over forty years at the University of Tennessee College of Veterinary Medicine, was my "go-to vet" for any medical issue that exceeded our ability to deal with at the local level. His seniority, coupled with an amazing breadth of knowledge and an innate love of pigs, made him a powerful resource for our sanctuary over the years. Dr. van Amstel pioneered some cutting-edge surgical techniques for our farmed pigs that few other veterinarians would even consider.

For a small sanctuary like ours, an active and strong board of directors is an absolute necessity. Our board members also became our biggest supporters, valued counselors, and very good friends. My heartfelt thanks to board members **Carl Lowe** and his wife **Teresa**, **Janet Baxter**, **Sue Levitt**, **Dina Brigish**, **Lana Hollenback**, and long-time intern **Matt Zeiser** for keeping me on the straight and narrow path for these many years.

To fellow sanctuary directors, **Ron** and **Lorelei Pulliam**, the late **Mary Ann Piatkowski**, **Elaine West**, **Carol Eiswald**, **Penny Yocum**, **Sue Parkinson**, **Sioux Robbins**, and a dozen or so more who served in the trenches with me during the early years when we were all struggling to survive and overcome the stigma of being the "red-headed stepchildren" of the farmed-animal rescue world—my hat is off to you for your dedication, help, and friendship over the years. Our collective motto back in the early days was: "We have done so much for so many with so little for so long that now we can do anything for anyone. The difficult, we do immediately; the impossible takes a little longer."

To the hundreds—or maybe thousands—of our supporters over the years, from those who gave hundreds of dollars each month to the equally important folks who contributed five to ten dollars monthly, know that without your financial help, there would have been no sanctuary and many thousands of pigs would have gone to their deaths. Each of you, regardless of the amount of your contribution, is a true partner alongside us in the sanctuary.

Special thanks and a deep expression of love to my daughter, **Jennifer Johnson**, and my granddaughter, **Grace Ann Gerrish**, for encouraging and believing in me over all these years, when the rest of the family either ignored me or treated me like the slightly crazy relative whom nobody talks about and who doesn't get invited to family gatherings. Of all the relatives on both sides of the family, you two "get it."

To **Tim Lewallen**, our "hired hand" and one of my best friends, who came to love the sanctuary and the animals as much as I did, who worked seven days a week all through the winter when I was sick and in the hospital to make sure that the animals were fed, watered, and cared for, who called me down when I got up on my high horse and kept me in check when I started to make an ass of myself: you and Stacey have my thanks, my love, and my respect.

To **Dick Hollenback**, who continues to show me, after all these years, what true friendship should be: You were always only a phone call away, always up for a road trip to rescue a pig or haul one to the vet school at UT. You helped me pry pigs out of mud holes and bury dead pigs and pitched in with a thousand other chores around the place for no more than a thank-you and a cup of coffee or a Diet Coke. I am proud to be considered a friend of yours; we are truly brothers from different mothers.

And finally, to my long-suffering wife **Laura**, who got me into the "pig business" and who has regretted it ever since: Thank you for your patience and understanding and your help when things got tough and I was overwhelmed. You always came through and never once let me or the animals down. You have sacrificed more than any woman should have to sacrifice and endured more than was ever right to ask you to endure. Over the years, you have shown me, through your words, deeds, and actions, what true love is. I hope that I have been worthy of your love.

This book is not an autobiography. It is a memoir. Simultaneously working twenty-four–hour shifts with the fire department and trying to run a rapidly growing pig sanctuary in the early years has taken its toll on both my body and my mind. As such, trying to accurately recall the names, faces, and stories of specific pigs among so many thousands after over thirty years has been excruciatingly difficult at times. And over time, among the hundreds of untold, almost daily miracles and tragedies that make up the life of any longtime sanctuary director, many details have grown fuzzy and indistinct in my head. For every story in this book, there are dozens and dozens not included because of a lack of space or because I could not recollect all of the facts sufficiently to tell them accurately. But the contents of the book are mine and mine alone. I bear sole responsibility for any errors or omissions.

Anita Krajnc

Every once in a while, a dog enters your life and changes everything! My journey of becoming a lifelong animal rights and climate action organizer began with Mr. Bean, a companion dog I adopted in 2010. Bean was the catalyst for the formation of Toronto Pig Save and Animal Save Movement. It's funny how a dog can put you in touch with stark realities in your own backyard.

On leisurely morning walks with Bean, I witnessed sad and scared pigs in slowly moving transport trucks in rush hour traffic, heading to a downtown slaughterhouse a mile from our home. This led to the formation of Toronto Pig Save and three vigils a week to bear witness to the pigs' suffering and exploitation. Taking action was also influenced by reading Leo Tolstoy, Mahatma Gandhi, Ramakrishna, Vivekananda, Saul Alinsky, Martin Luther King Jr., Cesar Chavez, Lois Gibbs and other inspirational community organizers and writers who organized in their local communities when there was an egregious injustice. The obvious thing to do was to come close and try to help the innocent, young pigs. Tolstoy said in his book *On Life*: "A person knows the life of other beings only through observation and only so does she know of their existence. She knows of the life of other beings only when she wishes to think of it."

The goal of helping as many people as possible bear witness to farmed animals at the front gates of slaughterhouses depended on staging frequent direct actions and employing a love-based community organizing approach. That goal was beginning to be realized when Animal Save Movement went global. Hundreds of chapters began holding vigils to shine a spotlight on the systemic violence animals were enduring. The activists around the world began organizing compassionate actions with their own cultural variations and diversity. In Mexico City, Valezca and her team climb roofless trucks to give water to the pigs and cows in front of slaughterhouses located in the dangerous outskirts of the city. In Colombia, Azul speaks of activists going arm in arm in one giant hug around the trucks calming the animals with short-lived human love. This book contains the insights of dozens of activists interviewed about their organizing experiences and rescue work. It details their courageous attempts to deal with the trauma of coming face to face with hell on Earth (slaughterhouses are like battlefields Tolstoy remarked) and how they begin to develop self-care and community care programs.

Reading Tolstoy continued to act as an anchor for me over the course of a decade of vigils. I distributed dozens of copies to other activists and re-read his favorite book, *A Calendar of Wisdom*. It grounded me between 2015-17 when I was tried for criminal mischief, interference with property (i.e., the pigs) after I had given water to thirsty pigs at a vigil. I'm very grateful for the support from the global animal rights community, my vegan lawyers James Silver and Gary Grill, Susan, my sister who was always by my side, Jo-Anne McArthur, Jane Velez-Mitchell, and PETA's president Ingrid Newkirk. I'd also like to thank Louise Jorgensen who organized the courthouse vigils, and the hundreds of other activists who came to the courthouse and Fearmans vigils.

After the Pig Trial, the focus on pig vigils, vegan outreach, celebrity engagement, and a handful of other tactics continued. But something was missing. We needed a campaign with targets, a greater repertoire of strategies and tactics, and timelines. Holding regular slaughterhouse vigils didn't suffice as a full-fledged campaign. I was painfully aware that we weren't re-enacting the full spectrum of practices prescribed by the movement leaders of the previous century.

Luckily, we discovered the Fossil Fuel Non-Proliferation Treaty when we met with its chair Tzeporah Berman in 2021. Later that same year, we created the Plant Based Treaty. We launched this global initiative by leveraging the global community of dedicated activists and groups that we set up around the world committed to bearing witness. We face runaway climate change if we don't have a revolutionary change in diets and plant-based food systems. We are facing the fight of our lives. It's all coming together though. . . from vigils to neighborhood block organizing, seed distribution, community gardens, vegan food giveaways, schools and city campaigns, and global advocacy.

My favorite people are animals and activists. A local vegan family emerged very quickly when we started holding weekly pig vigils in 2011. I want to thank every person who has gone to a slaughterhouse vigil and contributed to Toronto Pig Save and Animal Save Movement. There are too many to mention. When Animal Save Movement went global, we similarly developed into a global vegan family. I'm very grateful to the global core team and the networks of local organizers. I and so many have never experienced such unified, passionate, and talented global villages of vegan activist families.

When Regan Russell, a pioneer in animal rights activism, was tragically killed by a pig transport truck on June 19, 2020, time stopped. Activists and groups around the world came together in an act of solidarity. New and extant activists committed to holding the torch for Regan. Witnessing the outpouring of love and grief from so many around the world solidified my gratitude to the dedicated family of activists we have built.

I want to thank farm sanctuaries who have taken in animals rescued at slaughterhouse vigils. It was a fateful trip when Joanne O'Keefe suggested we go on a Big Pig Trip with Caroline Wong. When I stepped onto a farm sanctuary for the first time, I felt I was on a different planet which resembled paradise. The pigs, cows, turkeys, chickens, and other farm animals were respected and had names. There was a next level of appreciation and natural living at Richard Hoyle's The Pig Preserve. I saw Blue-Eyed Wilbur listen for minutes to the sounds of the forests. Farm and feral and potbellied pigs made friends with each other and formed groups of their choice as they foraged for food in the forests and grassy hills of Tennessee. There isn't anything Richard wouldn't do for the pigs, and he was far more concerned about creating the ideal homes for the pigs than looking after his own house. In fact, he and his wife's home had wide open doors to the animals that needed extra care and treatment. When I heard Richard speak of the pigs' language and gestures, and habits with endless love and respect, I encouraged him to tell his and the pigs' stories. I hope his approach spreads far and wide.

The manuscript would not be possible the way it turned out without the brilliant copy editing from Jennifer O'Toole. I also want to thank James O'Toole, Susan Krajnc, Kali, Mishel, Daisy, Joanne O'Keefe, and Nicola Harris. As well as, Liza Barkova, Pauline Lafosse, Emily Lavieri-Scull, and Hanh Nguyen from Lantern Publishing & Media. I tremendously appreciate Brian Normoyle, publisher at Lantern, for offering careful guidance and encouragement.

I dedicate the book to Mr. Bean, the pigs, and Regan Russell.

Foreword

Ingrid Newkirk

When it comes to having a central nervous system, and the ability to feel pain, hunger, and thirst, a rat is a pig is a dog is a boy

It was because of a pig that I was once thrown out of a Girl Scout jamboree. Earlier in the week, I had been driving along in southern Virginia and, seeing a sign by the road reading, "Piglets for sale," had stopped and bought one. She was about as big as a bicycle basket, a gorgeous shade of pink, and she squealed with delight at having her stomach rubbed or at anything edible.

That weekend, PETA was running a booth at the jamboree, so I took her along in an elevated toddler's crib. Our booth was set up in a big tent in a vast meadow, with food vendors on one side and someone selling coloring books on the other. I posted a sign that read, "Meet Harriet. Your bacon has a face."

The Girl Scouts who stopped by cooed over Harriet and promised there'd be no more bacon or ham in their homes if they had any say-so. A few even cried at the idea of Harriet losing her life. We discussed how weird and awful it was to see other living beings simply as food on the hoof.

The organizers got an earful from the girls' parents and troop leaders. This was too upsetting. We were too honest to be tolerated. They closed down the booth. The hot dog stand next to us remained, of course, the smell of flesh wafting through the fairgrounds. That heavy smell, once you came to realize what—or rather, who—it was, was enough to make you lose your lunch.

James "Jamie" Cromwell had a similar epiphany to that of the young Girl Scouts when he starred as Farmer Hoggett in the film *Babe*. He liked dancing with the animatronic pig, but the real piglets, so full of play and curiosity, troubled him. As a justice-oriented person, the fact that pigs like this would lose their lives for a ham sandwich didn't sit right with him, so he stopped eating them and all other animals as well.

My own "aha moment" occurred years before *Babe* was released. I was a law-enforcement officer in Maryland at the time, and my cases mostly involved physical violence to dogs and cats or occasionally to wildlife: a child setting a steel trap in the woods, a person shooting a raccoon with a crossbow.

One afternoon, I was called to a farm where the tenants had skipped out on the rent. The farmhouse wasn't the only thing the former tenants had abandoned. In the barn were the remains of animals who, unable to escape, had been left to starve. Only one piglet remained alive. I scooped him up in my arms and raced with him out to the pump. He was too weak to hold his head up, so I held it up for him as I dripped water into his mouth. He drank slowly and made little snuffling sounds of what seemed like immense gratitude. I radioed for someone to come to transport him to the veterinary clinic while I collected evidence from the house, the door of which swung open, trash and unopened mail littering the floors.

It was my last call of the day. I was feeling pretty somber because of what I had seen in that barn, and as I drove home, I was hoping hard that the piglet would recover.

It was getting dark, and I was hungry. I wondered what I had to eat. These were the days before microwave ovens, and with some relief, I remembered that I had taken a package of pork chops out of the freezer that morning, leaving them to defrost.

That's when the penny dropped. Here I was, determined to prosecute the people who had left a little pig to die of thirst. But was I so unimaginative that I couldn't picture what must have happened to the pig who was now in my fridge? The pig whose throat must surely have been slit after a terrifying journey in one of those tractor-trailer trucks you see on the highway. And all because I had paid the supermarket, which, in turn, had paid the slaughterhouse to do the dirty work. As Plutarch wrote, "But for the sake of a little mouthful of flesh, we deprive a soul of the sun and light, and of that proportion of life and time that it had been born into the world to enjoy." That's when I stopped eating pigs.

In 2017, I stood outside the Fearman slaughterhouse in Burlington, Canada with other many other vigil-keepers and protestors, hoping to prick the conscience of even one of the truck drivers ferrying doomed animals through the gates. That morning, Anita Kranj was on trial for daring to offer a few sips of water to other pigs, ones who had arrived there months earlier, when the weather had been so very hot for them and their journey had been long and frightening. I am not religious, but I thought of the parable of the Samaritan helping the injure man when "holier" people walked past him, uncaring, and how this "Good Samaritan" now risked jail for her act of mercy. When a doomed individual, already beaten down and panicked, looks up at you, it is hard not to do something for them. It is also hard not to cry—not only for their suffering, and, in this case, one's own helplessness, but at the injustice of a world in which pigs are viewed as property, not persons. And the cold realization that they have no rights and that many people take no responsibility.

The truck drivers did not have a change of heart, but I know a lot of young people who did. For some after I founded PETA, we took in animals from young Future Farmers of America and 4-H Club members. Parents called us because their sons or daughters could not bear to part with the pig, chicken, or lamb they had raised, named, looked after, and told their innermost secrets to. The children were supposed to learn that they could sell their friends for real dollars, but some of them did not want the object of their affection to end up in pieces at the butcher's. They rejected the lesson and saved their pals.

How like us are pigs? Well, so like us that in Borneo, cannibals have reported that human flesh tastes so similar to pig flesh that they called humans "the long pigs." Pondering pigs' other characteristics—that they, too, are emotional, social, curious, feeling individuals who value their lives—how difficult can it be to realize that we should treat them with compassion?

Philip Glass, when told that "it's sometimes hard to find non-leather shoes," reportedly said, "It's harder for the cow if you don't." When someone moans to me that it's hard to find a vegan alternative to bacon—although it actually requires only a modicum of effort, even in the most rural of areas—I say, "It's harder for the pig if you don't."

Ingrid Newkirk is president of People for the Ethical Treatment of Animals (PETA), the largest animal rights organization in the world, and founder of all PETA entities in Europe, Asia, the UK and the U.S. She is the author, among other books, of *Free the Animals: The Amazing True Story of the Animal Liberation Front in North America (30th Anniversary Edition)* published by Lantern Publishing & Media in 2022.

BEGINNINGS

1

A Pig Is Not a Funny-Looking Dog

LAURA ANN BROWN AND I MOVED into a house together in Stafford, Virginia, in 1987. As we unpacked our belongings, we learned about each other's eccentricities. What I noticed about Laura was her plethora of pig knickknacks—pig pictures, pig figurines, and so forth. By the time we were finished, the motif of the house could be described as "Urban Pig."

Laura told me that she had always wanted a pig for a pet.

I absolutely did not want a pig.

I had just left the Marine Corps and was working for the Fairfax County Fire and Rescue Department. Years ago, when I was a young lieutenant in the Marine Corps, it had been common for three or four neighbors to buy a pig in the spring to raise for slaughter in the fall, with the meat divided among the families.

That was what I knew about pigs—you ate them.

Our brick-and-mortar Stafford bungalow was on 1.5 acres of land, with woods on two sides. Though it was zoned agricultural, with cattle farms around us, none of our neighbors had livestock. A pet who might grow to be one thousand pounds was entirely impractical.

I felt safe.

What I hadn't counted on was that breeders would begin importing miniature Vietnamese potbelly pigs from Europe for zoos. In an unforeseen twist, movie stars like George Clooney began buying them as pets. When Laura upped her crusade for a pig, I would placate her by saying, "Wouldn't that be nice," while quietly rolling my eyes and changing the subject. These miniatures cost $30,000 to $40,000 a pair.

I still felt safe.

As the fad for potbelly pigs intensified, a few people in our area began to breed them. In the spring of 1988, Laura saw an ad about a breeder in the next town who was going out of business. He had one litter left that he was trying to get rid of. The cost per pig? $400.

Grudgingly, I said, "Yeah, okay, we'll take a look." I was really hoping she would forget it.

She didn't.

After I had worked my way through every reasonable excuse, we drove to the breeder's farm one Saturday afternoon on a trip billed as a "recon mission." The breeder showed us a black sow with about a dozen little black potbelly babies milling about in a large pen. I still hadn't come to terms with the fact that I would be leaving this farm $400 poorer and one pig richer.

I'd like to say we picked out one special little pig with whom we felt a moment of magical bonding: "There! That's our beautiful little angel."

Instead, as these squealing little black shapes scattered everywhere in their pen, the breeder chased after them with a fishing net. When he caught one, I heard myself say: "We'll take it!"

It was a purely emotional purchase, with not a logical thought in the whole process. I gave in to Laura because I figured that if I didn't, my life would be a living hell for at least the next seven months. Yet, in all honesty, something about these little pigs intrigued me.

Our pig fell asleep on Laura's lap on the way home, snuggled deeply in a blanket. I thought to myself: *We'll make this work.*

Our piglet was probably four weeks old and weighed maybe two pounds soaking wet. Breeders were saying that miniature pigs grew to be thirty to fifty pounds, but this pig's mom had looked much larger . . . an easy two hundred pounds by my reckoning. I dismissed this distressing thought because, as I have said, no logic went into this adventure. Until we arrived home, Laura and I didn't even know what sex we had purchased—female, as it turned out.

I was raised in an Irish Catholic family, and one of my mother's favorite old sayings was, "We're as Irish as Paddy Murphy's pig." We named our pig Patti Murphy, the female derivation. She was all black with a smushed-in face, straight ears, a pronounced potbelly, very short legs, and a straight tail. This, I would later learn, was typical of her Vietnamese "I" breed, one of maybe a dozen so-called miniature pig breeds found from India and Southeast Asia to New Zealand.

Laura and I already had three dogs, so I reassured myself: *We'll raise Patti Murphy as a funny-looking dog.*

It took about two hours for us to realize we were dealing with a creature dramatically different from a dog. This tiny pig was substantially more intelligent and definitely had a mind of her own. Unfazed by her new environment, Patti Murphy began nosing around, investigating her new world and her new family. She rolled canned goods off shelves. She opened the door to the refrigerator and tried to crawl inside to get at the food. She sniffed the dogs and walked away, the dogs sniffed her and walked away, each having agreed that the other did not exist.

Laura and I observed our Patti Murphy in fascination, trying to figure out what we had gotten ourselves into. We had no idea—none, whatsoever—about how we were going to deal with such an energetic little house pet.

Over the next couple of weeks, we learned that Patti expected us to play by her rules. You can control puppies by tapping their nose and with other mild but firm doses of corporal redirection, but Patti Murphy was totally unfazed by these methods. She would just stand her ground, look me in the eye, and holler right back at me in piglet indignation.

She could not be intimidated.

When Laura and I left Patti alone for an evening, she taught herself how to open the kitchen cabinet doors. She wrestled out canned goods, pasta, and a twenty-five–pound bag of baking flour, which she spread throughout the entire house before coating herself with the remainder. Months later, Laura and I were still vacuuming flour out of nooks and crannies.

Patti slept in our bed for the first few months . . . until she grew too big. Potty-training her was a simple, effortless process. Whenever she felt the need, she would go to the door and bump it with her snout. She had a few accidents early on, but her natural instinct was to go outside. She absolutely would not pee or poop anywhere near where she ate or slept—an early indication of the fastidiousness of pigs.

Since Laura and I didn't have any front-yard fencing, we broke Patti to a light, modified dog harness then took her out on a leash several times each day. She loved the outdoors, but she hated the harness, and she hated the leash. After she put on a few pounds, she would wear us out, dragging us everywhere, then scream her lungs out when we tried to take her back inside.

It became obvious to us that Patti needed her own out-of-doors time. Smart and fast as she was, we couldn't release her until we had piglet-proofed a space for her. After I had fenced our front yard, Patti walked around the perimeter a couple of times, sizing up her territory. She tasted some grass and decided she liked it. Sometimes she raced around, helter-skelter, making sudden turns, running in circles, rolling around in the grass, all the while squealing loudly with joy. We could see no logical reason for this other than her pure enjoyment of her freedom. We described this as "Patti doing her zoomies."

To give Patti access to both house and front yard, I checked out commercial pet doors. None were big enough for a pig. After examining their construction, I took down the back door into our laundry room, cut a Patti-sized hole in it, then manufactured a flap so she could be an indoor–outdoor pig. Since we still harnessed her out-of-doors, we worried about leaving her alone. If her harness caught on something, and she became frantic, we were afraid she might hurt herself. By the time Patti outgrew the biggest dog harness I could modify, I had expanded our front yard to give her more freedom.

As pig parents, Laura and I realized we would have to rethink many aspects of our lives. My job as a fireman/paramedic required me to work at least ten twenty-four–hour shifts a month, which left twenty days in which I could spend plenty of time with Patti. Laura was working nine-to-five as a nursing assistant. When I was on shift, she took over Patti's care and feeding.

I don't know what caused Laura to fall in love with pigs at an early age, but I've since met hundreds of seemingly normal people who, for reasons unknown, were as entranced with pigs as Laura. I still count my own conversion as one of the strangest, most bizarre events of my life. In about a month and a half, I was head over heels in love with Patti Murphy. She was such a devilish little creature—and so destructive. She would pull covers off the bed. She would wiggle into a pile of clothes and throw them all over. She investigated every nook and cranny in the house and could open any door she was able to reach. She rooted up carpet and linoleum and had an unnatural interest in pulling off strips of wallpaper. I don't know what it was, but she just got to me. I believe that having Patti also brought Laura and me closer together. Both of us had children from previous marriages and didn't want any more. Patti gave us someone we both could love.

I was amazed when Patti learned how to unlatch the gate on the chain-link fence that separated our front yard from our big pasture. I would go around to the back of the house after carefully locking it. Five minutes later, Patti would be grazing beside me, with the gate swinging free. Out of curiosity, I hid around the corner of the house to observe her.

By watching me, Patti had figured out that the little horseshoe latch had to be pushed up, and the gate pushed out. Using her snout, she pumped up the latch while using her shoulder to bump open the gate. Not only had she figured out how the gate worked, but she'd figured out how—with her snout and her shoulder—to do what I did with my hands.

Some researchers say that pigs have the intellectual capacity and maturity of a four-year-old human child. After watching the cunning ways in which Patti solved problems, I would say she was more like a five- to six-year-old.

We never could trust Patti where gates were concerned. My solution was to keep building bigger and better fences and sturdier, more complex gates. This accommodation arose out of one of my first realizations: Patti had a primary fear of being confined. We were dealing with a creature who, in her natural world, was a "prey animal," with no defense other than escape or evasion. To constrain or to otherwise curtail Patti was to terrify her.

At first, we fed Patti the same food as we did our dogs. After reading up on pigs, we switched to commercial pig food. Common sense told us that wasn't good for her either. As a product developed for factory farming, it had three purposes: to grow pigs to be as big as possible, as quickly as possible, and as cheaply as possible, for the meat market.

Laura and I wanted to feed Patti for optimum, lean health and longevity.

We learned about Mazuri, a subsidiary of Purina, which produces food for exotic animals—literally every species from aardvarks to zebras. They made three kinds of miniature pig food in twenty-five–pound bags—for youths, for active adults, and for elders. Patti stayed on Mazuri for about ten years. By then, we had so many pigs we could design our own food and have it manufactured.

Patti's hearing was only a little better than a dog's. She was a bit near-sighted, the more so the closer objects were to her snout, but her nose was phenomenal—far superior to that of any animal I've ever encountered. In autumn, I would sometimes scatter acorns in the front yard, and she would scramble from one to another like a machine. She could smell out an acorn a hundred yards away and follow a trail of acorns forever.

Patti was a dedicated rooter and grazer. She would scoop out divot-sized holes in the soft earth, looking for tender grass roots and the occasional grub, all of which she devoured eagerly. Though we worried that she might ingest a poisonous plant or mushroom, we soon realized that she had an innate sense of which ones to eat and which ones to leave alone. She would strip a rosebush of every rose but ignore an adjacent azalea bush. She would wander through a stand of new mushrooms, eating some and rejecting others. The more we observed Patti's behavior, the better we felt about her ability to fend for herself. Like all parents, we struggled to find that right balance between allowing her too much freedom and being overprotective.

As Patti rounded out from a gangly piglet to a traditionally shaped mature potbelly "I" pig, we grew concerned that she hadn't been spayed. If you adopt a female dog or cat, you take her to the vet and get that done. Laura and I called all over Virginia and Maryland, trying to find a vet who would work on miniature pigs, but couldn't find any. Not one.

This lack of readily available veterinary care became serious when Patti was about a year old. After Laura and I had rearranged our bedroom furniture, we lifted Patti up onto our high fourposter for the night. While we were still asleep, Patti jumped off the bed. Instead of landing on a safe spot as usual, she hit her head on a dresser and knocked herself out.

On hearing a loud thump, Laura and I jumped out of bed. Patti seemed groggy and a bit unsteady on her hooves, but otherwise okay. We hoisted her up into bed and went back to sleep. About an hour and a half later, Patti had a full-blown Grand Mal seizure. She also vomited and defecated. We cleaned her up, then took her back to bed. Forty minutes later, Patti had more seizures.

Since this was happening on New Year's Eve, Laura and I spent the pre-dawn hours of New Year's Day tearing through the phonebook, looking for a vet who would see Patti. Finally, we contacted a twenty-four–hour exotic-animal clinic in Springfield, thirty-two miles away. Their on-duty vet warned us over the phone, "I don't know much about miniature pigs, but I'll have a look at her."

We raced Patti up Interstate 95 to the clinic, where the vet took x-rays, then poked and prodded for a while. Finally, he said: "I think she's just had a concussion. I'll give her a shot to keep her brain from swelling. Either she'll recover or she won't. We'll just have to wait and see."

Though his words weren't very encouraging, Patti did get progressively better that New Year's Day. She continued to improve over the next three or four days.

That whole experience severely frightened Laura and me. Losing Patti at that point would have been like losing a child. We also felt responsible because we had rearranged the furniture without making Patti aware of it.

We redoubled our efforts to find a permanent vet and finally located Dr. Carole Nicholson, a recent vet-school grad who lived nearby and who worked out of her home. I told her, "You probably don't know much about pigs, but we need someone who will at least try to take care of our Patti."

Dr. Nicholson had had only about two weeks of training on pigs, mostly geared to those in a factory-farm environment; however, she possessed good common sense, and her expertise, combined with my years of experience as a paramedic, proved a good combination.

We never did get Patti spayed. Because she was no longer a piglet, that would have meant a complete hysterectomy, and we weren't prepared to take that risk on her behalf.

As Patti grew into her second year, she became ever more mischievous. Since she wasn't allowed to go into the bathroom, she would slip in there when we weren't looking, then hide behind the door. By the time we looked behind the door, she would have already slipped out. When we turned our backs, she would slip in again.

Patti also liked to flip over her water bowl, then lie down in the puddle and splash. It took us a while to realize that she wasn't being contrary. She was telling us something. Since pigs don't sweat, they can't use evaporative cooling in hot weather like other mammals do. Without water to cool themselves, they can easily suffer a heat stroke, often fatal.

Bathing, we discovered, is also a very personal activity for pigs.

Patti found a soft spot in our front yard. It was just a small depression with a bit of rainwater, *but, boy, did she ever go to work on it.* I was amazed at how much earth that shovel-shaped snout could send flying. She was like a little four-legged bulldozer as she sculpted the hole so it was the right size for her body, even molding it to her shape. Watching her was like watching an artist at work. Patti then flopped down into the hole, turned over onto one side, and kicked her feet. She turned over onto her other side and kicked some more. She sat up on her butt, then scrunched back and forth in the mud. She was so happy. It was such a joy to watch—indescribable!

Mud protects pigs from sunburn. It also protects them from biting insects and parasites, but I believe the main reason Patti wallowed in mud was because it felt so good.

Over the next month or two, Patti sculpted her mud hole bigger and bigger and bigger, with a shallow end and a deep end. I finally had to put timber around it to keep her from occupying half the front yard. Of course, when it was time for her to come inside, we had to give her a bath or at least wipe her as clean as we could, which was a procedure Patti detested.

When Patti was age two, she went through a six-month period when Laura and I wondered if we had made a mistake in adopting her. That was a very rocky time for us. With children, this is known as the "Terrible Twos," but it felt more like we were dealing with a rebellious teenager. After doing what she wasn't supposed to do, Patti would dare us to stop her. She also became quite snappish. She was losing her first set of teeth and growing her adult teeth, including her upper and lower tusks, which are pig canines. She began nipping at Laura's legs with those sharp teeth, while also trying to knock her down.

Laura was naturally perplexed, "What's going on with Patti?"

From my frantic research on pig behavior, I learned that pigs, when raised in nature, are considered babies for the first eighteen months to two years, during which they have no social standing in their herd and are allowed to run wild. The whole herd raises them, like a village raising a human child. Then, at around eighteen months, these pigs undergo a metamorphosis, at which point they are no longer babies and must figure out their social standing: *Who is the alpha pig in my herd? Who is the next in rank?* And so on down the chain of command, until they identify the lowest pig in the hierarchy. That's the pig they challenge first in their quest to establish their own rank.

Patti's "herd" had three dogs, plus Laura and me. Patti had ignored the dogs until she reached this watershed in her maturing life. I had already noticed—without understanding why—that Patti had bullied her way up through our three dogs before attacking Laura.

Patti had begun by running after, and snapping at, tiny black Grover, our most timid dog. After he backed down a few times, Patti went after our next least aggressive dog, an Irish Setter named Red Dawg. When she found him sleeping, she would push him awake with her snout. When he didn't respond with aggression, she took that as a win, then tried that same strategy on our top dog, a Corgi named Little Bit, with the same passive result. Patti was lucky that our dogs were very laid-back. If they had retaliated, Patti would have received the worst of it because she was defenseless. Since our dogs weren't using Patti's playbook, they didn't know why this silly little pig was suddenly in their faces. They ignored her, which must have allowed Patti to believe that she was three steps up on her herd's dominance ladder, clearing her way to take on Laura.

This was unacceptable, but what to do about it?

First, I had to remind myself that we were not dealing with a dog or cat. We were dealing with a creature significantly more intelligent, significantly more emotional, and significantly less "domesticated."

Though Patti did not respond to corporal punishment, I had noticed how carefully she watched my body language, and how good she was at picking up on my moods and attitudes. I could tell her two or three times not to do something without any effect. If I turned and put my hands on my hips, she would take me more seriously. If I also took a step towards her, that had a much stronger meaning. If I put my hand on Patti's shoulder and pushed, the way a pig would use their snout on an opponent's shoulder, that had the strongest meaning of all.

Through trial and error, I had been learning to communicate with Patti through body language, and also by mimicking any vocalizations I could associate with specific behaviors. I told Laura: "If Patti attacks you, don't give ground. Correct her in the same way as a more dominant pig. Put your hand on Patti's shoulder, and push just enough to make her take a step or two back. Use your sternest voice, but don't scream or yell at her."

Working with Patti so that she accepted me as the "alpha pig," and Laura as second-in-command, took about six months. We had to be patient and willing, because Patti would sometimes sneak up on us, take a nip, then run like hell. She was not striking out in anger. She was doing something innate, required of her by her species. She was finding her place as a member of our herd, during a period that was probably as scary and confusing to her as it was to us.

Once the pecking order was established, with the three dogs having lost the contest they didn't know they'd entered, Patti seemed perfectly comfortable with her dominance ranking. All her aggressive behavior melted away, and she once again became our sweet little pig.

Laura and I have always believed that you should never have just one of a species. You need at least two of the same kind to keep them stable and happy. If we had one dog, we needed to have two dogs. If we had one cat, we needed to have two cats. If we had one pig, we needed to have two pigs. This seemed especially true for Patti because she was more socially needy than any dog or cat.

To find a pal for Patti, Laura and I faced a serious hurdle: miniature pigs were still ridiculously expensive. Once again, fate stepped in. On my days off from the Fairfax Fire and Rescue Department, I volunteered as a firefighter and paramedic with our local Rockhill Volunteer Fire Department. One Saturday, while I was hanging out at the Rockhill firehouse, another volunteer told me that he had found a young pig running loose in his yard. A breeder across from his house was going out of business, and apparently this pig had escaped. The volunteer had caught the pig with a fishing net and was intending to fatten him for meat.

I asked, "Where is this pig?"

"In the back of my van."

When I took a look, I found a frightened, squealing little black pig, about twenty pounds and six months old.

I said, "I'm going to take this load off your hands."

Since I was offering payment without the nuisance of having to raise the pig, coupled with the fact that I was the fire chief, the young volunteer quickly agreed. So now I had a companion for Patti. This little pig, who had been intended for someone's dinner, was also our first official rescue.

Pig #2 was the same "I" breed as Patti, except he had three white feet and was a genetically bigger male. We called him Damien Gilhoolie, and he was such a destructive little rascal that he made Patti, now developing into a more sedate lady, look like a saint. *Oh, man, he was a tornado.* Laura and I would say, over and over, "Boy, that Damien sure is a stinker." After a while, this was shortened to the nickname "Stinky Pig," then "The Stinker," and then just "Stinky." That was the name that stuck to Pig #2 for the rest of his life.

Stinky's exuberance was our first introduction to pig testosterone. He was a little boar, sexually active at about three to four months and interested in only two things: eating and making baby pigs. Now, little intact boars are obnoxious, headstrong, and randy porcine dynamos of energy and personality. Getting Stinky neutered became our number one priority. Luckily, removing pig testicles is a fairly simple and straightforward surgery; however, for pigs in the real world, this operation is typically undertaken without any sedation or pain management. It took some doing, over and over in the years to come, to convince many vets that for all surgical procedures on our pigs, sedation and pain management were vitally important.

With the introduction of Stinky into our herd, the competition for dominance began all over again. For about two weeks, Patti would go after him every time she saw him, as if she were going to tear him apart. He was a lot smaller and a whole lot faster, because he also had fear on his side. He had no desire to confront someone that big and that ferocious. It was tempting for us to intervene, but we realized that Patti was just doing what she was supposed to do. She wasn't really going to hurt Stinky. She was just letting him know that she was the alpha pig. If we broke that up, we wouldn't be solving anything. The friction between them would go unresolved, with confrontations happening when we weren't around to supervise.

After these initial battles died down, Patti and Stinky went through a period of about a month when they studiously pretended each other did not exist. Next, they would stand five feet apart and grunt and growl, and snap their jaws at each other before disdainfully walking away. After that, we would occasionally find Patti and Stinky lying down together. Finally, they became totally bonded. They ate side by side. They were never more than five or six feet apart.

Laura and I did not find Pig #3. Pig #3 found us.

Some college fraternity kids over on the eastern shore of Maryland had purchased a pig as a prank. To hide her from their housemother, they'd locked the pig in a closet. Now, they didn't know what to do with her. After agreeing to take the pig, Laura and I drove to the college to pick her up. She was a white European Royal Dandy, with a longer silhouette than the typical potbelly and blue/black markings on her butt. We named her Shamrock.

The fire service has a saying: "If you do something once, it's unique and innovative. If you do it twice, it's standard operating procedure. If you do it a third time, it becomes a tradition."

After Laura and I accepted Shamrock, every couple of weeks, our phone would ring, and someone would say, "Hey, I understand you take in pigs." Then, they would pour out some sob story that would persuade Laura and me to reply, "Well, yeah, maybe just one more."

One of our more challenging early rescues involved a little pig we used to see grazing beside a big quarter horse on a horse farm near us. When we heard from neighbors that the pig had been attacked by dogs, we introduced ourselves to the family to offer our help. They told us that a wild pack had broken into their pasture, causing the horses to bolt in one direction and the little pig to run in the other. The dog pack had jumped the pig and mauled him pretty badly. He was saved only when his friend, the quarter horse, returned to drive off the dogs.

This family had put the little pig into an unused horse stall, full of old bedding and manure, to let nature take its course, with no thought of veterinary care. When Laura and I saw him, he was almost dead, covered with hideous wounds and bite marks, especially around his head, face, and neck. After his owners agreed to let us take him, we called our vet, Dr. Carole Nicholson, to patch him up and to start him on antibiotics.

For two or three weeks, it was touch and go whether this pig was going to survive. Even after he seemed to improve, he retained a badly damaged ear, which the antibiotics didn't appear to be helping. Early one morning, when I went to check on him, I found maggots on the wound. I called Carole, very concerned.

She told me: "The little pig's infection has gone down into his ear canal, where the antibiotics don't seem to be working. If we can't control it, the infection will eat into his skull and kill him. Though maggots are repulsive little creatures, they will only eat the necrotic flesh. Leave them be, and let's see what happens."

Every time I checked on the little pig, his wound was so nasty. However, after about five days, the maggots had disappeared. Two days after that, his ear fell off, leaving a clean hole with no sign of infection. He was a small fellow, probably under a hundred pounds, and he looked like a wizened pirate with a scar across his eye and one missing ear. Laura and I had started calling him "Little Old Man," and that name stuck. Little Old Man lived fifteen more years with us before dying of natural causes.

During the mid- to late nineties, the fad for potbellies began to wane. In the eighteen months after our rescue of Stinky, we acquired over two dozen other miniatures. Some we were able to place in good local homes—a labor-intensive option, requiring us to educate prospective "pig parents," with follow-up visits to make sure all was well. We also had to work with county officials in regard to zoning—another lesson to add to our rapidly expanding knowledge of pigs. Some jurisdictions legally classified miniatures as "swine" or farm animals, while others recognized them as companion animals. Many jurisdictions simply hadn't dealt with the issue before and were in no hurry to do so.

Often, we felt like we were running an out-of-control nursery school for hyperactive four-legged kids. Our first three—Patti, Stinky, and Shamrock—remained our primary house companions. In the morning, we'd let them out the gate into the pasture. At feeding time, they would come in, then stay for the night.

Patti continued to be the alpha pig—the queen of the herd, to whom the others deferred. When Stinky was two, he challenged her a couple of times, just as she had acted out at that age. Though he was now much bigger than Patti, she went after him as in days of old, smartly putting him in his place. These skirmishes didn't upset us as much as did the original ones. Patti and Stinky were now functioning within a group, the dynamics of which we, as pig parents, were beginning to understand. The two of them were no longer dependent on each other for all of their social needs. Stinky had his buddies, and so did Patti, especially her pal Shamrock.

With the burgeoning of our herd, it wasn't just Patti's relationship with Stinky that was evolving. Now that Patti's social needs were being met by other pigs, her relationship with Laura and me was also changing. Even when Laura and I were home, Patti would spend less and less time with us and more with the other pigs. At first, we were resentful: "She doesn't love us as much as she used to." But this wasn't about love. Patti didn't *need* us as much. Given the foreshortening of a pig's life, this was like a human child leaving the parental nest to explore life as an individual. It was her choice. It also lifted pressure from us. We no longer had to feel guilty about leaving Patti alone, because she was no longer alone—she had a real pig family.

Laura and I also made a formal change in our relationship. We were married on October 23, 1993. My fire department chaplain and Laura's father, who was a preacher, married us in his little Baptist church in Stafford, right up the road from where we lived. It was a dual ceremony, with just a few close friends in attendance. Laura's son David was there, along with my middle daughter Jennifer and my first grandson Christopher, still in a bassinet.

We did try to wrangle an invitation for Patti Murphy, but the church was adamant in its refusal.

When casting an eye to our future, Laura and I found ourselves at a crossroads. Though our land, just outside of Washington, DC, was designated as an agricultural property, housing divisions were sprouting up all around us, and our neighbors weren't too happy with our ever-growing collection of pigs. Like most people, they thought of them as dirty, smelly, noisy creatures, when they are basically clean, odor-free, and quiet. The endless demands of misinformed neighbors and local politicians alike became the springboard for a question that would shape much of the rest of our lives: *Should we divest ourselves of all our pigs except for Patti and Stinky, or should we purchase more land and learn how to look after our herd the right way?*

In the end, that decision made itself for us. We knew most of our pigs weren't adoptable. Those who were might end up in situations as bad as the ones from which we'd rescued them. While casting about for a solution, we discovered the term *sanctuary*: "a place of refuge or safety; a nature reserve." We had heard of sanctuaries for elephants and primates and birds, but did sanctuaries for pigs even exist?

Whether they did or not, Laura and I decided we would create one, with me as the director. What were my qualifications for the job? Quite by accident, and certainly without intention, I was already running one.

2

Chance . . . or Fate?

THAT DAY IN THE SUMMER of 1986 had seemed like any other.

Because of my passion for firefighting, I could often be found at the Rockhill Volunteer Fire Department on my days off from my professional work with Fairfax County. Since this local department was not as busy as the urban ones, I would sometimes wander half a mile up the road to an old country store, which doubled as a fledgling small-engine repair shop. It was run by David Blackman, a retired marine gunnery sergeant whom everyone called "Mr. B." Between customers, he and I would drink coffee together and tell war stories, as two "old salts" are wont to do. When Dave's small-engine business picked up enough for him to need another mechanic, I started working for him part-time.

One hot day in August, I was at the shop, repairing a tractor, when an attractive woman stopped by to have a flat on her truck fixed. She was driving a 1960s Chevy pickup, which this very personable young woman told me she had restored and rebuilt herself. I plugged her tire as she requested, and she drove off, leaving a favorable impression on me.

Now, one thing a volunteer fire department must do is raise money. Ours held a bingo game every week to which the whole community would show up. We firemen took turns on a roster, so once a month I had to work the bingo crew.

At one of these games, I noticed this woman whose truck tire I had fixed, and we chatted. I discovered that she was recently out of a marriage, as was I. She had a young son by a previous marriage, and I had three daughters, but none of our children were living with us. She was staying with her parents and working for her dad as a mechanic. As an aside, she was one of the best car mechanics I've ever come across.

We struck up a relationship.

I was frankly surprised that such an attractive woman, twelve years younger, would be willing to put up with me. We seemed well matched because neither of us had a need to be the dominant figure in the relationship. Some things I was good at, so she deferred to me. Some things she was good at, so I deferred to her. Both of us had an extremely low tolerance for drama, and I also appreciated her honesty. This amazing woman was willing to give me the time and space that I needed to be a firefighter and a paramedic. I was willing to support whatever she needed to do, because she did bounce around for a while. After quitting her father's shop, she was a waitress at a Cracker Barrel restaurant, and then she went on to become a nursing assistant, launching a career that would take her to a top hospital administrative role. I admired her as a self-made woman.

Her name was Laura Ann Brown.

Nothing in my life before Laura could have made me believe that I would spend thirty-six years rescuing and nurturing pigs.

I was born in the Panama Canal Zone in 1948 to a devout Irish Catholic family. Since the Canal Zone—a ten-by-fifty–mile area surrounding the canal—was a US territory, my two sisters and I automatically became US citizens, like both our parents.

My father, William Warner Edward Hoyle, had migrated to the Canal Zone after attending the city fire academy in Providence, Rhode Island. This was during the Depression when, in order to land a secure government job, you had to have political pull, which my father did not have. Unable to fulfill his dream of becoming a fireman, he drifted through several jobs before members of his family told him: "Your uncle is a retired sea captain working in Panama as a canal pilot. There are jobs in Panama. Why don't you go there?"

My father took that advice, and the retired sea captain did find him work as an agent for various shipping companies. He would meet ships as they came into harbor, prepare a list of the supplies each one needed, then make arrangements to acquire those materials. After a few years, my father became a US customs agent for the Panama Canal government, eventually rising to chief of customs for the Panama Canal.

I'm not sure how, or why, my mother, Rita Marie McEvoy, percolated from the States down to Panama. I believe she went to visit relatives and just stayed for the same reason as my father: jobs were available. As an incentive, American citizens were paid a bonus for working outside of the United States, and I believe they also received tax advantages. Though each of my parents had relocated for financial security, the decision to stay in the Panama Canal Zone with its perpetual summers, as compared to New England with its formidable winters, might not have been too difficult to make.

Both my parents' families had migrated to the US from Ireland several generations past. My mother's people were from West Cork, which made them dyed-in-the-wool Catholics. My father's family came from Northern Ireland and England, which made them—God forbid!— Protestants. Their decision to marry stirred up religious hate on both sides of the family that lasted a number of years.

Back then, in order to marry a Catholic, the non-Catholic partner had to agree that any children would be raised in the Church, to which my dad acquiesced. Over the years, my mother's incessant harping, and the unrelenting zeal of a couple of very aggressive Jesuit priests, wore my dad down to the point where he converted to Catholicism. Whether this was just his way of getting my mother and the two Jesuits off his butt or a true conversion of faith, I do not know. I assume it was a combination of the two.

My mother, who was a force to be reckoned with, went to her grave knowing that I was destined to be a priest. She was as convinced of this as only an Irish mother could be. In preparation for my eventual entry into the seminary, she decreed that I would become an altar boy at Saint Mary's, our local parish church. Altar boy school (a.k.a. Boot Camp) was run by the local Maryknoll Sisters. Several nights each week, we were schooled in the various services of the church and forced to memorize a tremendous amount of Latin. The Maryknoll Sisters—a benign and fairly easygoing bunch of nuns—soon handed us off to the Sisters of Mercy, a much more militant order we students quickly dubbed the "Sisters of No Mercy." Looking back, I believe the drill instructors at Marine Boot Camp could have taken lessons from these nuns.

On my graduation from altar boy school, my mother saw that my name was represented in virtually every service the church offered. This included 8 a.m. Sunday Family Mass, Sunday noon High Mass, the six-days-a-week 6 a.m. mass, 7 p.m. Monday Novenas, and the Stations of the Cross every Friday night throughout Lent. Then there were the ever-present wedding masses, funeral masses, and masses on every Holy Day of Obligation.

For years, my long-suffering father would wake me up at 5 a.m., drive me up the hill to Saint Mary's Church, then push me out into the dark parking lot so I could serve the 6 a.m. mass, which concluded at 7 a.m. This left me just enough time to race down the hill to our house, grab my books, then make it to school at 8 a.m. Admittedly, I cut a rather dashing and saintly picture in my altar boy tunic and cassock, but since I was spending more time in church than were many of the priests, the services took on the air of a military detail rather than a religious experience. The day I received a pardoning of my sentence (by convincing my mother that my high-school studies, and my life as a teenager, could not co-exist with my altar boy duties) remains one of the high points of my life.

I never felt close to my mother, the family's disciplinarian. I believe that I was very spoiled initially, because I was the miracle baby. Almost eleven years before I was born, my parents had had a son who died shortly after birth. The doctors told my mother that she would never conceive again, but eleven years later, I—the baby who wasn't supposed to be—came along. I also had two younger sisters, Virginia and Cathy, born at roughly three-year intervals, but the fact that I was the firstborn, and a son, had special meaning for my Irish Catholic family.

I was very close to my dad. He was that stereotypical, good old, hard-drinking Irishman who let the vagaries of life roll off his back. My relationship with my mother always seemed to be a series of running skirmishes. Though my dad ran interference for me when he could, he loved my mother dearly and never directly crossed her on my behalf.

My parents had considered enrolling me in parochial school, but after several intense discussions with the nuns who ran the school, it was "strongly suggested" that I was definitely a candidate for public school . . . a blow my mother had a hard time accepting.

I actually enjoyed school. Though the academic program was never difficult for me, I found out early that if you bust your butt to receive a very good report card, your parents will expect that every time. I managed to set my bar high enough to please my parents but low enough that I didn't have to kill myself working.

Looking back, I can see how much effort my parents invested in trying to see that I turned out to be a good man. Not only did our family always go to Sunday Mass, but they instilled in us obedience to the Ten Commandments throughout the week. They also taught us never to consider ourselves above hard, manual labor, implanting in me a good work ethic. My father was not only one of the finest men I have ever met, he was also one of the handiest—skills he passed on to me, he did along with his values. The Panama Canal Zone had a distinct caste system separating different races. My parents never subscribed to that, so I grew up with a lack of prejudice.

Though I couldn't have asked for better role models, my parents were well into their forties when I was born. By the time I was a teenager, they were old by current standards. They gave me far too much freedom, which I religiously, enthusiastically, and joyfully abused every chance I got.

I was in trouble all of the time—stunts that today would probably put me in jail. Back then, the Canal Zone was a very insular community in which everybody knew everybody else. It wasn't unusual, on a Friday night, for my dad to receive a phone call from the local police, saying: "We've got your kid here at the station for drinking beer and shooting at stop signs. If you want to pick him up, we won't book him."

This was in the sixties, when both parents and police expected teenage boys to be spirited and a bit rebellious, but we knew the limits to our rowdiness, and neither I nor my cohorts crossed the line into serious trouble with the law. In most cases, we feared the wrath of our parents more than that of the police.

I was never a sports fan. The Canal Zone had only two teams, one on the Atlantic side and one on the Pacific side, without tournaments to rally everyone into that *rah rah* frenzy. Though Panama was too damn hot for vigorous sports, I enjoyed year-round hiking and camping, fishing, hunting, and snorkeling. On my twelfth birthday, my dad took me to a local NRA-sanctioned rifle range and enrolled me in the Junior National Rifle Association. I became an excellent competitive shooter and an avid hunter, mostly of small game.

We had a family dog named Corky, because of his little corkscrew tail. I was as close to him as any kid would be to their first dog, but it never occurred to me that I would devote much of my life to caring for animals, or that I would ever see anything wrong with hunting and killing.

The Panama Canal Zone, with its government amenities, provided such an easy, idyllic life that many kids who grew up there had no desire to leave. On graduation from high school, they would gravitate to jobs with the Panama Canal, in a society so institutionalized that it was almost like a well-maintained prison.

In high school, I joined the Army Reserve Officer Training Corps, which inspired my desire to enlist in the military. My parents—anticipating that the United States would eventually give the Canal Zone back to Panama—impressed on my sisters and me that we should choose an American occupation, go to the United States after high school, and never return to the Canal Zone. On my graduation in 1966, I decided on The Citadel, a military college in Charleston, South Carolina. Though my mother still hoped I would apply to a Catholic university in preparation for entering the seminary, my dad knew priesthood was not my calling and secretly encouraged me to make a different choice.

During my high-school years, I had developed an intense dislike for the army, which felt too big and amorphous. I became increasingly attracted to the United States Marine Corps as a smaller, more elite group of fighting men, with much higher standards and a long, glorious tradition of service to our country. Once enrolled at The Citadel, I signed up for the Marine Corps Reserves as well as for the Marine Corps Commissioning Program, run jointly by the college and the Marine Corps.

I loved the physical challenge and the discipline of military training offered by both The Citadel and the Marines, but unfortunately, I had to leave The Citadel after my junior year.

As a favor to one of my roommates, I had taken his cousin Anne, who was a year ahead of me at Furman University in Greenville, SC, to a couple of big military events such as Homecoming. In the course of all the partying, drinking, dancing, and other typical college shenanigans, Anne and I were intimate, and she became pregnant. Abortion was not an option for me, being Irish Catholic. As a devout Protestant, Anne felt the same way, and abortion wasn't legal in the sixties in any event. If you were raised as I was, you do the right thing. You marry the girl.

Since cadets at The Citadel must remain single, I had to tell the commandant of cadets about my "problem" and my plans to marry. After checking my record, he said, "You'll have to leave, but we'll let you finish the semester."

This was a terrible blow. I had an impeccable record at The Citadel and was the top choice to be a company commander in my senior year, a position for which I had worked hard.

It was an emotional and difficult period, requiring me to do a whole lot of "growing up" in a very short time.

Since I desperately wanted to be married in the Church, I took my dilemma to our Catholic chaplain. After listening closely to my story, the good Father dashed my hopes with his refusal. Not only was my wife-to-be "with child," but she was not a Catholic. My home parish was overseas, and the church in Charleston had no ready way to verify that I had been baptized and confirmed in the Catholic Church.

I was deeply offended. Hadn't I paid my dues as a loyal Catholic? Now, in my hour of need, when I was facing marriage to a wife I barely knew and fatherhood for which I was totally unprepared, my church had turned its back on me. Though that was the logic of a nineteen-year-old, I have not stepped inside of a Catholic church since that day.

Anne and I were married in Atlanta, Georgia, by a Baptist preacher who was a very good friend of hers. The wedding party consisted of Anne's college roommate, the preacher and his wife. Since we were both still in college, there was no honeymoon. I dropped Anne and her roommate off in Greenville, SC, then headed south on I-26 for Charleston before my forty-eight–hour pass expired. I was still a Citadel Cadet . . . for a few more weeks.

When my parents received news of the quickie wedding, along with the information that Anne was a Protestant, they created an instant replay of their parents' fury over their own mixed marriage. Anne's father, who was a deacon in the Baptist Church, and her mother were equally scandalized that their daughter had been made pregnant by an Irish Catholic and that now she was marrying him. Our families disowned us. Though matters did improve on the birth of their grandchild, my relationship with both sets of parents was permanently damaged.

After my untimely departure from The Citadel, I decided to take a bachelor of arts degree at the University of South Carolina. I had entered The Citadel as a lance corporal in the Marine Corps Reserves, and I had been promoted to the rank of corporal. Now, the Marine Corps told me that if I completed my bachelor's degree in four years with a C average, I would be promoted to second lieutenant. If not, I would be sent directly into active duty in Viet Nam as a corporal.

The University of South Carolina's requirements for graduation turned out to be a fraction of The Citadel's. By completing my junior year at The Citadel, I had accumulated so many credits that I needed only one semester at South Carolina to graduate. In 1969, I received my bachelor's degree and was commissioned as a second lieutenant in the United States Marine Corps Reserves. It was a muted and somber ceremony, very anticlimactic to the graduation and commissioning that I had envisioned just a few months before.

After I completed the basic officer course in Quantico, VA, I was ordered to Camp Lejeune, NC, for twelve weeks of training as a combat engineer. I was then sent to Southeast Asia, where I spent a year as the platoon commander of a combat engineer platoon, attached to a Marine infantry battalion. It was a tough year, physically and emotionally. It changed me a great deal.

Even today, I rarely talk about my war experiences. I do believe that many of the things I saw and did during my overseas tour helped shape my future. Hundreds of thousands of young men had to find ways to deal with this period in their lives. Many learned how to cope. Many did not. And we did not return home to a joyful, welcoming nation. Mostly, we were reviled. That denigrating reception made our readjustment to normal life even more difficult. The Viet Nam War is still a complicated issue in my mind. Even today, I hold a deep-seated grudge against a nation that treated its returning veterans so shabbily.

Back in the States, I spent several fulfilling years as Marine company commander, with over 250 young Marines in my company. Then, after a year or so of running a specialized school, teaching the use of explosives and landmines, I found myself behind a desk, growing increasingly dissatisfied as I pushed papers from one side to the other. Eleven years in the Marine Corps, during which I became a captain, had taught me that the higher I rose in the ranks, the less time I would spend as a Marine and the more time I would spend as office staff. I was a good Marine officer, but I was a poor politician and an even worse glorified office clerk.

While I was trying to decide whether to stay or to leave the Marine Corps, I was offered an enticement: to attend the Naval Postgraduate School in Monterey, California. I took the bait and graduated in 1979 with a master's degree, majoring in organizational development and human resources management and minoring in education. What I hadn't fully understood in the contract's fine print was that I would then be required to spend four years at Marine Corps headquarters in Washington as payback—my version of hell.

With those four years at last behind me, Anne and I bought a house in Rockhill, Stafford County, in northern Virginia, just around the corner from the local volunteer fire department. By then, we had three daughters—Lee, born in 1969; Jennifer, in 1973; Erin Colleen, on New Year's Day in 1979. As soon as we had unpacked the last box, I was banging on the fire department's door, offering to volunteer.

I loved being a fireman. It presented me with the challenge of dealing with a constant series of crises and emergencies, which was what had originally stimulated me in the Marine Corps. Firefighting became my passion. My dad had always regretted that he couldn't be a fireman. Somehow, he passed that dream on to me.

In 1984, after seventeen years of marriage, I walked out on Anne and my three daughters. Since I believed the failure of our marriage fell mostly on my shoulders, I gave up so much in the divorce settlement that I was left poor and homeless. For the next few years, I lived at several volunteer firehouses in exchange for running emergency calls day and night, seven days a week—not conducive to a balanced life, but I was able to pour heart and soul into being the best firefighter I could be.

One thing about the Irish: they're good at carrying grudges. My mom and dad, who had shunned me because I had married a Protestant, now ostracized me because I had divorced a Protestant. From the day that Anne and I separated, I saw my parents only once. When my dad died, no one bothered to tell me until well after the funeral. I heard only because a brother-in-law was kind enough to call me. Same with my mother's death. I didn't find out about that either, not until well after the fact. Today, I exchange occasional emails and Christmas cards with my sisters, both of whom live in Oklahoma, but that's it.

During the divorce, I had explained to my three daughters that I would not insist on visitation because I thought it unjust to force them to spend every other weekend with dad. I also told them that I would find a place within ten miles of their home so they could visit me any time they chose. I would always have a bedroom for them, and I would be only a phone call away if they needed me.

My ex-wife, Anne, did a magnificent job raising our girls. I waited in the wings, trying not to interfere in the lives of my daughters, who believed me to be an evil ogre. I kept my promise to be available by buying a small house in our local community, and by trying to help them out on the rare occasions when they got into trouble as teenagers.

Thankfully, after many years, I now have a fantastic relationship with my middle daughter Jennifer, who is a special-education teacher and the mother of four of my five grandchildren. Jennifer and her kids are one of the happiest parts of my life. I'm deeply saddened that I haven't achieved the same rapport with my other two daughters.

I finally decided to leave the Marine Corps in 1985, after attaining the rank of major. At thirty-seven, I was too young to settle into a corporate job, but too old to qualify for the job I really wanted—to be a full-time fireman. Most fire departments wouldn't hire anyone in their thirties—that is, until a court challenge declared this age discrimination. I raced to the Fairfax Fire and Rescue Department in Fairfax County, Virginia, scored well on all its tests, including the physical ones, and was hired.

Now, the FFRD required every firefighter to be a basic Emergency Medical Technician. The department sent me to George Washington University, which had the contract to train paramedics. After eight months, I passed the university's exam, and then the state exam, and then the national exam, qualifying me as a full-fledged paramedic.

Now, I had the best of two worlds: I could put in my twenty-four–hour shifts as a career firefighter and paramedic while spending my days off volunteering with the local Rockhill fire department and the local volunteer rescue squad. My work as a paramedic challenged me emotionally and intellectually, while firefighting was more of a physical challenge. Every day presented me with different emergencies that exhilarated me. I did everything a big-city fireman would do. I pulled people out of burning buildings. I pulled bodies out of burning buildings. I cut people out of cars. I probably did thousands of CPRs both as a paramedic and as a firefighter. Since my rural community was about forty-five miles from the closest hospital, and the roads were treacherous in wintertime, I delivered a lot of babies, either in homes or at roadside. When I counted it up, it came to twenty-six babies in thirty years. Sadly, I also attended a few firefighter funerals.

After Laura Anne Brown and I had dated for a year, we decided to move in together. Our three-bedroom bungalow was on a Stafford County country road, about a quarter mile from the Rockhill volunteer firehouse. As the two of us furnished our new home, I noticed a plethora of pig knickknacks—pig paintings, pig figurines, and so forth.

Laura began telling me that she had always wanted a pig for a pet.

I absolutely did not want a pig.

THE SANCTUARY

3

A Good Kind of Crazy

WITH TWENTY-FIVE MINIATURE PIGS NOW in residence on our 1.5-acre Stafford County property, Laura and I were ready to create a sanctuary. We loved Patti Murphy, and Stinky, and Shamrock, and all the others, with their quirky, inquisitive personalities. We wanted to give them more than we could in our currently limited space.

In our search for role models, we were amazed to discover one of this country's first dedicated potbelly-pig sanctuaries, located in southern West Virginia, less than three hours up the road. It was called PIGS, A Sanctuary and run by partners Dale Riffle and Jim Brewer.

Dale and Jim became caught up in the pig-rescuing business in the same circumstantial way as we had. While living in an apartment in urban Maryland, they had bought a cute miniature pig for a companion animal, not realizing how life-changing this commitment would be. When they were told by local officials that they couldn't keep a pig in their fancy apartment, they got rid of the apartment instead of the pig. Dale and Jim then put all their resources into the purchase of five acres in West Virginia, where they created PIGS. By the time Laura and I discovered them, they had been in operation for almost six years and had already accumulated around four hundred miniature pigs.

On a visit to PIGS, Laura and I liked what we saw and volunteered to spend every free day working there. During the next couple of years, I'd say to Laura: "Okay, you've got a long weekend, and I've got three days off. Let's go to the sanctuary." For special PIGS projects, we hired our veterinarian, Carole Nicholson, as a pig-sitter for our gang, allowing us to spend a week to ten days of our vacation there.

Those two years were like an internship in which Laura and I learned a great deal about pigs, as well as about running a sanctuary. Some of the practices at PIGS we thought were very smart. A few we wanted to do differently. We also saw the dedication and the investment that it took to run a grassroots sanctuary. When Laura and I had first volunteered, Dale had asked me, "Why do you want to operate a pig sanctuary?"

I had replied, "Because I love our pigs."

"Loving pigs is not enough."

It took those two years at PIGS, plus a decade of experience, to really appreciate what Dale was telling us. As owners of a sanctuary, we would be committing ourselves to a monastic lifestyle in which every day revolved around feeding schedules and the tending of sick pigs. We would have very few people in our life. The money we saved on frivolous activities, like evenings out or vacations, would go into supporting our pigs.

Even while Laura and I were interning, we had been looking for a suitable piece of land for our sanctuary. At last, we found a seventeen-acre property that we thought would suit our needs in Culpeper, in the foothills of Virginia's Blue Ridge Mountains. The house—single-story with a basement—was in reasonably good shape; however, we delayed the closing for ninety days to build our first pasture and to put in a barn in preparation for moving our herd.

Our plan was to create a layout in which two people with full-time jobs could effectively feed and care for a large number of animals. When completed, our house would be an administrative hub, with five or six rectangular or trapezoidal pastures extending out like the spokes of a wheel. Each pasture—ranging from two to five acres—would include some grass and some woods, and enough small barns to house its pigs. Both the feed barn and the tool shop would be next to our house.

In 1998, Laura and I transported our twenty-five pigs in a stock trailer the fifty miles from Stafford County to Culpeper County. This move would mean a major work commute for both of us. Laura would be driving seventy-five miles, five days a week, to Charlottesville, where she was now a hospital administrator. I would be driving seventy-five miles, ten to twelve times a month, to Fairfax. I would also have to give up both my volunteer firefighting and my volunteer work as a paramedic.

In the days, weeks, and years that followed, this would become our routine: On one of my twenty-four–hour shifts for the Fairfax fire department, I would get up at 3 a.m. in order to feed our special-needs potbellies, then leave for my 7 a.m. shift. Laura would get up at 5 a.m. to feed the rest of the potbellies, then head to work by seven. When she arrived home at night, she would have to do all the evening feeding. She would also do the feeding the next morning before I arrived home at around 9 or 10 a.m. On my off days, I would do all of the feeding, the rest of the animal care, and the maintenance on the barns, equipment, and house.

We incorporated our sanctuary under the name Mini-Pigs, Inc. and acquired our 501C3 status, which designates it as a tax-exempt, nonprofit organization. We also appointed a small board of directors, with our vet, Dr. Tom Massie Jr., as our medical advisor. At its maximum, our board consisted of eight directors, each with a different area of expertise.

After buying the Culpeper property with our own savings, Laura and I set up Mini-Pigs so that we would pay the mortgage and the insurance and absorb all of the administrative costs. We would also provide the farm equipment and pay for its maintenance and repair. That meant donor funds would be used only for the direct care of the pigs—feed, veterinary bills, some fencing, and barns.

Our first donations came from people who wanted us to take their pigs. I'd say: "Look, we'll be glad to help you out, but your pig may live another ten to fifteen years. Would you consider giving $20 to $50 a month to sponsor her?"

As the saying goes, "a gentleman's agreement presupposes that both parties are indeed gentlemen, and will be true to their word." In about

50 percent of these cases, the money came in for only a few months, then disappeared. Obviously, these people knew that we weren't going to abandon their pigs just because they did not honor their commitment. On the other hand, some donors upped their monthly amounts, then continued regular payments even after their pigs had passed away.

Tragically, we soon learned that an estimated 90 to 95 percent of miniature pigs obtained as companions would die, or be abandoned or surrendered to a sanctuary or shelter, before their second birthday. The US Department of Agriculture's classification of miniature pigs as livestock meant that they are not protected under the Animal Welfare Act. Breeders are not required to have a USDA license to buy and sell them as companions, as they would be for a protected species. This made it easy for the unscrupulous to latch onto the miniature pig trend strictly for profit, flood the market with piglets—often inbred or sickly—then drop out to catch the next fad.

As we also discovered, the majority of breeders deliberately misled prospective guardians by assuring them their companion pigs would remain under fifty pounds and would instantly love them with the unconditional devotion of a dog. But pigs are not pink dogs, and they don't necessarily enjoy modeling a stylish wardrobe like the pigs who turn up on Instagram. Keeping a pig, in comparison to a dog or a cat, involves fairly substantial lifestyle changes. The majority of uninformed guardians soon found themselves overwhelmed, and since breeders steadfastly refused to take back the pigs, they were left without good options, for which the former paid the price.

The main reason why people abandon their pigs is because of size. They purchase an animal called "miniature," "micro," or "teacup" without understanding that the piglet they can hold in their hands now may, in four to five years, grow to be anywhere from 100 to 350 pounds, depending on pedigree and diet.

Snigglers was a typical victim of this kind of cruel abandonment. Her family enjoyed her when she was a baby, but when she wasn't so cute anymore, they didn't want her. They relegated her to hanging around the yard, where they fed her dog food and unsuitable leftovers. As she grew fatter, folds of flesh covered both of her eyes so she couldn't see. Then, of course, she'd bump into things, including people. Some in the family seemed to take pleasure in kicking her, screaming at her, throwing things at her, so whenever she sensed someone nearby, she would lash out and snap. She was a real fighter, constantly in a state of confusion because she was blind.

After rescuing Snigglers from her life of hell, Laura and I turned her into a pet project, determined to show her that we loved her. As an older female not in the best physical shape, she lost weight very slowly. Since she also enjoyed terrorizing other pigs, people, dogs, and anyone else who ventured within range of her jaws, she spent a lot of time by herself. However—if I can jump ahead six years—Snigglers slowly lost enough weight so she could see. She also learned to trust Laura and me sufficiently to let us give her belly rubs. She even learned to tolerate two or three other yard pigs, though she was never close with any of them. Let's call them acquaintances rather than friends.

Elvis, a ten-year-old barrow (a male castrated before puberty), came to us with the opposite problem to Snigglers'. His guardian thought she could defeat genetics and keep him "tiny" by giving him only one quarter of a cup of feed each day—a strange and ironic contrast, since she herself weighed 250-plus pounds.

Elvis had all the earmarks of a chronically malnourished pig—oversized head, poor skin, no hair, emaciated body, and chronic arthritis. Though his guardian still wanted to keep him, she was forced to surrender him after she lost a zoning battle. We knew Elvis could not survive the rest of the winter at the sanctuary in his condition. We told the woman we would take him in the spring if she could increase his weight by a minimum of thirty pounds.

When she returned with Elvis, he was still pitifully thin, but no longer emaciated. On accepting him, we were determined to put another thirty to forty pounds on him before the next winter.

Elvis understandably had a tough time learning to be a yard pig. His adjustment was made even more difficult by his former guardian's insistence on visiting him every other day to shower him with affection, increasing his sense of abandonment when she left yet again. One day, when she arrived unannounced, she was aghast to find Elvis covered in mud. He had just discovered the pig joy of digging a mud hole then wallowing in it. On a second visit, she observed scratches on Elvis's flanks and an ear tick from tangling—one time only!—with our ornery Snigglers. She was so upset, I thought I would have to call 911. I told her, in no uncertain terms, that she could not return until Elvis had made his transition from a household companion animal to a real little pig, with piggy habits and desires.

Without interference, Elvis made remarkable progress. Soon he was making friends, eating ravenously, and getting into all sorts of pig mischief.

A more hidden reason for our acquiring many of our potbellies was domestic abuse.

While we were still in Stafford, a woman appealed to us to take her companion pig Daisy. When the woman was out of the house, her abusive husband would vent his fury on poor Daisy by beating her up. The woman was preparing to leave her husband and to relocate to a place that didn't allow pigs. She believed that if she left Daisy with her husband, he would kill her.

The day we opened the trailer door for Daisy and watched her experience sunlight and safety for the first time was truly joyous. This scene, which we would see repeated many times, became a strong stimulus for our wanting to stay in the pig-rescue business.

The reason for Daisy's surrender proved all too typical. An estimated 25 percent of the potbellies who came to us at Mini-Pigs were surrogate victims of human domestic abuse, some arriving with serious injuries. This shocked us into realizing how much violence must be happening in American homes.

Li'l Richard was referred to us by Jennette Ferro of New Orleans, a woman well known in the pig community for her work as an animal rescuer. She contacted us to ask if we would take a small, older boar who was one of the ninety-plus she had single-handedly saved from death. Out of all those precious and pitiful pigs, Li'l Richard had stolen her heart. Though he was in pretty sad shape when she took him in, she nursed him back to health. Now, with our permission, she planned to drive him to us in Virginia, along with three rescued females whom she was taking to a different sanctuary.

Li'l Richard turned out to be a magnificent old boy with the sweetest disposition, who looked a bit like a miniature buffalo. It was easy to see why Jeannette had fallen under his spell, just as we soon did.

For a couple of days, we penned Li'l Richard by himself under the house, where he immediately made friends with two of our rabbits. Due to his advanced age, we had decided to let him be one of our yard pigs, rather than putting him in a pasture.

At first, Li'l Richard was not too confident about his new surroundings. He tentatively explored the yard and met some of our pasture pigs by snout-sniffing through the fence. He tested our Virginia grass and found it to his liking. He walked away from several potential confrontations with other yard pigs, until he came up against Snigglers (still our work in progress). In an epic but short-lived battle of the titans, Li'l Richard actually ran Snigglers off, which boosted his pig ego to no end.

Life was about to become even sweeter for Li'l Richard. While strutting around savoring his victory, he chanced across Cherie, an older pasture pig who liked to wander into the yard during the day before returning to her pasture at night. Cherie was in heat. In spite of the fact that Li'l Richard had been neutered months before, he still had some inkling of what boars were supposed to do. He mounted Cherie for a brief sexual tryst, before we herded him back into his pen for the night. As I closed the gate, I could imagine him kicked back in a pile of fresh hay, serenely contemplating his two victories and feeling pretty certain that his new life here at Mini-Pigs was worth the long trip from New Orleans.

"Pig" arrived at our sanctuary just before dark one spring evening, without an invitation or even a warning. A guy I'd never seen before drove up in a pickup truck, then informed me: "This pig that I've got in my truck is too mean to keep any longer. Either you take him, or I'm going to kill him."

Since his truck was shaking on its springs as a large and angry creature squealed, groaned, kicked, and banged, I had no reason to disbelieve his description. Still, a pig was a pig, and the mission of our sanctuary was to rescue pigs.

I agreed to take whoever was in this guy's truck.

Though I was concerned how we were going to unload this new arrival, I shouldn't have worried. As soon as the truck's tailgate was dropped, I saw a slobbering, snorting gray behemoth clear the truck by about ten feet, then disappear at a gallop into the woods. As he rocketed past me, I caught a fleeting glimpse of a long tusk, a foaming mouth, and one large eye.

When I say this pig was big, I am not talking about one of your pet potbellies who may be a few pounds too plump because of snacks slipped to them by doting guardians. I'm talking about maybe 450 pounds of fury. As he disappeared into the twilight, I swear I could see trees shaking as he crashed and bellowed his way through the woods. By now, every pig on the property was up and squealing, the dogs were howling, and I was quite sure our assorted cats were at the top of every convenient tree.

The driver quickly pumped my hand in a gesture of sincere relief and left a massive $15 donation. He told me our new inhabitant's name was Pig, then wisely got the hell off our property before I changed my mind.

Slowly, peace returned to the sanctuary, along with the darkness of night.

Early the next morning, I fed all the other pigs before I had the courage to step foot in what I now considered "The Beast's" pasture. Since Pig had not come forward to be fed, I carried a small bucket of feed into the deep woods, hoping it would buy me time to escape if Pig decided to ambush me.

It wasn't hard to find him. All I had to do was follow the trail of broken saplings and smashed vegetation that Pig had left in his wake. With some of that debris, he had made himself a nest. As I approached, he stood up and grunted at me. I was about to make a break for the nearest fence when I saw something that stopped me in my tracks. This big fellow's tusks had grown all the way back till they were embedded in his jowls. They had created two huge, gaping, oval-shaped wounds that lacerated his cheeks with every movement of his mouth.

Pig must have been in agony.

Suddenly, he was no longer "The Beast." He was just an oversized, poorly treated pig in great pain—a condition he had obviously been enduring for some time. I could see the suffering in his eyes and hear it in his voice. Pig was scared and hurting.

I left him a pan of feed, then called Dr. Tom Massie Jr. On my insistence that this was an emergency, he agreed to come to the sanctuary early that afternoon.

Dr. Tom was a busy man. I could tell when he arrived that he resented being paged for an "emergency" tusk-trimming. This is typically a routine chore, done when business is slow. I would set the pig on his butt, then hold him while Tom sawed the tusks with a "giggly wire" and rounded them off with a Dremel tool. Tusks are similar to ivory—tough but quite brittle. Since the roots go deep, care must be taken not to trim too close to the gum line, but the procedure itself is painless. For the bigger pigs, we may have to administer a mild sedative to restrain them, but in no way does a typical tusk-trimming qualify as an "emergency."

Tom lost his resentment as soon as he saw Pig.

We kicked around several containment options, all of which involved my risking my life to corner Pig long enough for Tom to stick a needle in his butt. Finally, in desperation, I suggested using Tom's compressed gas-powered dart gun to shoot a three-inch steel needle filled with anesthesia into Pig. Though we had not as yet field-tested this method, Pig seemed like the perfect candidate.

As Tom readied the artillery, we discussed how large a dose he should shoot into Pig. Tom favored a mild one that could fit in a 3 cc dart. Since life to me is sweet and precious, I suggested a much heftier dose, using a 5 cc dart. Tom wasn't comfortable with that for our first use of this technique, so we settled on the smaller dose.

Stalking and shooting was the easy part—for Tom, at least. It required me to offer my body as bait so Tom could sneak up behind Pig and shoot him in the butt. Afterwards, we retreated to the relative safety of the fence for about fifteen minutes, while the combination of sedatives we had devised worked its magic.

We found Pig sitting on the side of a wooded hill—groggy, but by no means sedated. Given the load of drugs Pig now had on board, Tom stated categorically that he would not be able to move. His plan was to throw a lasso over Pig to keep him from rolling down the hill while we worked on his tusks, trimmed his hooves, and vaccinated and dewormed him.

Over the years, I've learned that I should listen to the quiet nagging doubts that enter my mind when knowledgeable people make decisions that just don't sound right. Pig sat perfectly still as we walked up to him. He didn't even move when Tom dropped the loop of the lasso over his head. However, when it touched his front legs, Pig took off down the hill like a runaway locomotive, with Tom on the end of the lasso.

Now, Tom weighed maybe 120 pounds with a flatiron in each pocket, whereas Pig was easily 450. I tried to grab Tom, now surfing through brush and briars on his stomach at an alarming speed. Just as I caught up, Pig slammed on his air brakes. I overshot Tom but crashed into Pig, who then knocked me into the briars before taking off in an entirely different direction. I grabbed both Tom and the end of the rope as they shot by in a storm of leaves, twigs, branches, and briars. With my 250 pounds added to the rope, Pig was having a much harder time pulling. I managed to loop a small stump with my end, bringing Pig down in a heap. While Tom tied off the rope, I threw myself onto our thrashing Pig.

As soon as I landed, Pig quit struggling, then simply lay there as sweet as you please. Within ten minutes, during which Pig stayed perfectly still, we managed to do everything we had planned, including dressing Pig's facial wounds. When we released him, he stood up and walked off. Within thirty minutes, he was eating and drinking as if he had spent all his life tranquilly at the sanctuary.

Tom and I retired to the back of his truck to critique the event, to repair torn clothing, and to tend to our accumulation of minor cuts, scrapes, and lacerations. Our patient was uninjured, but we were a mess.

I changed Pig's name to Jack Duggan. He subsequently took up life in the backwoods pasture. That whole winter long, he steadfastly refused to move into a barn, a dogloo, a doghouse, a calf hutch, or any other shelter I dragged out to his pasture to entice him. He barely came out to eat with the other pigs. He lost weight, which he needed to do, but continued to drop pounds long after reaching his optimum.

Finally, with much trepidation, I lured Big Jack into the yard where I could monitor him. I did this despite my concern that his presence might not bode well for the other pigs, for ourselves, and for visitors.

I could not have been more wrong. Big Jack turned out to be a teddy bear who greeted every visitor at the road, who would pester us to scratch his ears and give him belly rubs. He would actually twist around our feet like a large cat, complaining loudly in ecstasy as we scratched and rubbed him. The only thing ever wrong with this gentle boy had been severe pain, plus a lack of pig companionship and human love.

The Jack Duggan episode was a clear example of how crazy Laura and I had been to take on the risk of creating a sanctuary, but it was a good kind of crazy. Working with pigs was a peaceful antidote to all that I had witnessed in my Marine Corps career, along with the tragedies I still encountered daily in my work as a firefighter and paramedic. I loved the pigs. Their innocent response to both pain and joy and their insatiable love of life were a tonic.

4

Pigs Don't Do Gratitude

It is a typical February morning in the year 2000, on one of my days off from the Fairfax fire department.

4 a.m. I awaken to the sound of sleet and freezing rain, and the infernal clash of the wind chimes that I hung for Laura last summer. I fight Sugar Pup, one of our six indoor dogs, for my overalls. He's sleeping on them, and doesn't like to be disturbed.

4:30 a.m. I head for the kitchen to make coffee, tripping over the "feral" cats Laura has allowed into the house, now twisting around my feet in hopes of spurring on breakfast. Though Mini-Pigs specializes in pigs, as that name makes clear, we were persuaded to rescue a colony of forty-seven cats. Though they were labeled "feral," the lure of warmth has turned most of them into purring occupiers of every comfy surface in the house. Now I must wade through a minefield of refrigerator magnets and knickknacks that they used, all last night, to play field hockey. Hard on the feet. *All God's creatures*, I remind myself.

I make the coffee, knowing it is too strong and Laura will give me hell for it.

5 a.m. I go into my office where Gus the Betta fish, a ten-year resident, now swims in his bowl with a large aquatic plant. Gus lives in constant fear of the cats, although they never actually threaten him. I sit down at my computer and begin to wade through some of the emails that have accumulated while I was firefighting. The DELETE button is hot when I am done, leaving about fifty requiring answers.

6 a.m. Laura is up, preparing for work and complaining about the coffee. After a quick and slippery dart outside, I unfreeze the door of her truck. *Oh boy, it's cold and nasty out there.*

7 a.m. Laura leaves for work. I finish my emails, mostly thanking donors and following up with vets, then I check out a couple of pig-related websites. I have two more cups of coffee, feed and water the cats, then chase as many of them outdoors as I can.

I go downstairs into the basement to check on Noel and her babies. Noel is a young, silver spotted girl who came to us near term, then delivered seven healthy piglets on Christmas Eve. She was one of four pigs—three females and a male—on sale at a local slaughterhouse for $2. When we received a pleading call from the woman who had rescued them, our first reaction was to say no. Our pastures were full. Our barns—even the three new ones—were full. Every quarantine pen and temporary hold pen housed at least one pig waiting to be moved into a pasture come spring. Two more pigs were on the way—a large female, now in Florida under threat of death, and a tiny baby from Tennessee, saved by a dog-rescue sanctuary that couldn't keep her. Yet when we received that call about Noel and her three companions, we knew deep down that we would somehow find room and the funds for "just these few." After a couple of weeks, we did.

We neutered the boar, Peanut. Although he weighed only about thirty-five pounds, he had been a very busy little boar while in the pen with three intact females. We also moved Noel into a stall in our basement—a former den, where she could warmly snuggle into bales of fresh straw to deliver her babies. When it seemed she would soon farrow, we checked on her every few minutes. Noel let us lie down with her in the straw to rub and talk to her—a very trusting and loving pig.

Each Christmas seems to bring a new rescue. During our first Christmas here, we received Petunia and Little Rosebud—the only surviving piglets from several litters, totaling around thirty-five, born to wild sows on a nearby farm. Both needed to be bottle-raised, but are now living in a pasture with Floral, their surrogate mom.

These rescues drive home the traditional meaning of Christmas much deeper for Laura and me. While much of the world wallows in the excesses of presents, parties, and decorations, saving a desperate animal, then providing her with shelter and love, seems to more closely approximate what the original Christmas was all about.

Now, as I fetch water for Noel's little family, I take pleasure in their soothing, contented grunts. Except for a stranger's kindness, this whole family would have been cruelly slaughtered. Noel is an excellent mother. She is also good about letting us mess with her babies. In spring, after she has raised and weaned them, we will introduce her to her new pasture home, where she will never again be cold, afraid, or hungry. Her babies will be moved to an outdoor pen under the deck, ready for adoption into good homes. If that doesn't happen, they, too, will spend their lives here, simply being pigs.

While engrossed in thoughts of Noel, our Christmas miracle, I accidentally leave the door of her pen ajar, letting the piglets loose. *God, those little dooterbutts are fast.* Though corralling them leaves me sweaty and winded, they regard this as fun before returning to their mom for a morning snack. At least this delays my departure into that wicked winter weather, which I tasted while unfreezing Laura's truck door.

8 a.m. I layer on clothes, preparing for the worst. We've held steady to six ill-mannered dogs, known collectively as "The Hooch Puppies," who provide the sanctuary's early-warning system for visitors as well as keep all the ferocious squirrels and rabbits at bay. The Hooch Puppies swirl around me as I negotiate the wooden ramp designed to let Patti, Stinky, and Shamrock into the house. With all this churning, I lose my balance and fall flat. I say many bad words, which the dogs don't understand, so they jump on me, licking enthusiastically.

8 a.m. to 10 a.m. January and February have been rough here at the sanctuary. Our mild winter weather was suddenly replaced by a series of snow and ice storms. Though the snow is beautiful, it makes life a lot tougher for both nonhuman animals and humans. Routine upkeep becomes difficult to accomplish, so when the snow and ice finally melt, a sea of mud remains. Tasks that must be accomplished in spite of the weather may now take two to three hours instead of just one. As a case in point, I wrestle with five sections of a garden hose that I drained last night then laid in a neat row. The dogs have dragged them all over the yard, so they didn't drain properly. Now all are frozen to the ground, along with their quick-release connections. It takes half an hour to get water flowing through them . . . with the dogs helping, of course.

I slip and slide from pasture to pasture. Each has a minimum of seven water containers that must be tugged loose from the frozen ground, then stomped on to loosen the ice inside. It comes out in giant cubes, allowing me to refill the containers. Since the hose's connections are still partially frozen, they separate several times, soaking me from head to foot. The water freezes on my clothes, glasses, and hat, so now I am walking stiff-legged, encased in ice.

At each pasture, the pigs run out, drink, pee, and poop, then hurry back into their cozy barns. I keep an eye out for split hooves, sprains, or otherwise damaged feet, along with dragging bellies, to ensure there is no frostbite. The older pigs, like Elvis and Li'l Richard, have the worst time negotiating the ice and snow.

Pneumonia and other respiratory problems worry us at this time of year, but luckily we have had no cases among our seventy-plus animals. Our pigs aren't happy about the weather or about being confined to their barns, but other than their being a bit short-tempered from cabin fever, all is well. Humans and pigs alike are hoping for an early spring.

The only pigs who seem to tolerate this awful weather are our Ossabaw Island "teenagers," fittingly named after the Rascals of early movie fame: Spanky, Alfalfa, Buckwheat, and Stymie. They actually enjoy sliding down the hill in the snow. Pigs of this breed evolved on Ossabaw Island off the coast of Georgia, where they grew wild. Because of inbreeding, they suffer from insular dwarfism as well as a low-grade form of diabetes, which resulted in their being used for human diabetes research. The bad weather and low temperatures don't seem to bother them at all. They can be seen in the worst snows, cavorting and playing, while our potbellies only come out to eat, drink, pee, and poop before ducking back into their barns.

I take a shortcut across one of the mud holes left over from last summer. Since the ice is not as thick as I expected, I break through into the muck. When I try to walk, my left boot pulls off. Instinctively, I balance myself on my right foot, like an ungainly giant crane, before putting down my stockinged foot, so now it is mucky as well as frozen.

10 a.m. to 12 p.m. I finish the watering, then drag back the garden hose, disconnect and drain it. I reach for the portable phone that I tucked into my pocket. It isn't there. In spring, the pigs will find it, along with several others, which they will gleefully disassemble.

Time for feeding. I dish up a small bucket of dog food, a smaller bucket of rabbit food, and yet another pail of goat food. Yes, we also have a pygmy goat named Rosie. Given all the evidence around her, she quite sensibly believes she is a pig. She takes a special mix of bran, Metamucil, and mineral oil.

During this nasty weather, I have been feeding our pigs using our tractor in four-wheel drive, with the feed buckets in the front-end loader. In winter, our pigs' calories must be increased to help them stay warm. The pigs hear me coming. A group gathers—hungry, cold, and irritable—on the other side of the gate, blocking my entrance. I set down my buckets so I can push back. I win, because they want what they know is in my buckets.

Once inside, I am mobbed by hungry pigs. Petunia and Cherie, two sows in heat, attack my legs, then try to "horny-pig" me. We neuter the males as soon as they arrive at the sanctuary. It's an easy "outside" job, requiring only mild sedation and a couple of injections of lidocaine for pain relief. Usually, they're up and walking around in an hour or so. Spaying the females requires a complete hysterectomy, which not many vets feel comfortable doing. The health and age of a sow also have to be considered, along with the cost—upwards of $1,000 as opposed to $50 for a male. As a result, each pasture has at least two extremely amorous females in heat, eager to mate with any intact male—pig or human. For their purposes, I will do. They start by rubbing their heads and slobbering, then proceed to nipping and—yes!—attempts at a forced mounting. I resist with vigor.

I make it to the troughs without spilling my buckets. The troughs are full of ice and snow so must be cleaned out. I can't put down the buckets because the pigs will upset them, causing mayhem. I trudge back to the gates and hang them on hooks, while the pigs grow ever more restless.

I finally get the troughs cleaned then filled while dodging a barrage of pigs, fast-moving like bowling balls. That's because a pig is interested not only in what is in her trough but also in what is in her neighbor's trough. The pigs jet from one trough to another, afraid someone is getting something they don't have. I try to count them to make sure everyone is present.

As I exit this pasture, Snigglers—yes, that old malcontent!—bites the back of my legs. This is to inform me that she has finished her bowl and wants more. I throw her half a cup of feed, then make my escape.

I deliver a bowl of bran with all the fixin's to one of our special-needs seniors, then kneel in the muck to beg her to eat while staving off the other pigs, who would like to help her out. I am nipped in the butt by Shamrock and head-butted by a couple of others.

There are ten troughs in the Big Pig Pasture. I slip and slide from one to another, dropping three cups of feed in each bowl. Molly plants herself behind my legs. Harrigan, Mulligan, and O'Flynn push me from the front so I fall backward over Molly, spilling the feed. This cunning attack was too well choreographed to have been an accident. I make yet another escape, leaving the herd to fight over the spill.

Now I feed Pokémon in his doghouse, which he never leaves. The only way I know he exists is that twice a day, he puts his empty bowl at the door for me to fill. When I bang on the roof, I hear grunting. If I put my hand inside his door without a full bowl, he bites it. It's never a serious bite, just the kind that says, "I'm not trying to hurt you, but I want you to know that I don't want you in my life." I suppose he comes out at night, drinks, pees, and poops. Once a month, I run him out of his doghouse just to see that he's okay.

We don't know enough about Pokémon's background to understand why he is such a reticent creature. One of our bigger pigs used to enjoy sticking his head into Pokémon's doghouse, which scared the hell out of him. When the head got stuck, we had to pick up the whole doghouse and shake it with Pokémon inside. It struck me as funny, but Pokémon never came around to seeing it that way.

I slip and slide down to the bottom of the pasture. Along the way, I feed Sebastian, who has a place all to himself. Sebastian is a small fellow probably about twelve years old. When he came to us, he was emaciated and covered with untreated mange mites. We gave him his own pen in the Big Pig Pasture so he could enjoy the other pigs behind the safety of the hog panel fence. Sebastian has gained weight and is no longer crippled. We suspect that he endured years of abuse before arriving here. He has learned to like his belly rubbed, but today he grumps while I clean and fill his dish.

Sebastian grumps a lot, but he is a sweetheart under his gruff exterior.

Finally, I herd Dolly into her pen and sit with her while she eats. If I don't, the other pigs will steal her food. She daintily picks up pellets one by one out of her bowl while my feet freeze.

12 p.m. to 2 p.m. Back at the house, I shed layers of frozen and soggy clothes, have lunch, then check my phone messages.

2 p.m. to 3:30 p.m. After suiting up again, I coax the tractor to start, load a dozen bales of fresh hay, then head for the barns in four-wheel drive. The tractor gets stuck, adding half an hour to my routine. For the past two months, we have been supplementing pig feed with alfalfa hay. This year, in Virginia, it is not only scarce but also expensive, thanks to summer's drought.

Every time I open a barn door to put in fresh hay, I am greeted with a chorus of angry squeals, grunts, and groans: *Shut the damned door! It's drafty in here.*

I slog from barn to barn, putting out fresh hay and administering to the pigs whom I identified during feeding as needing attention. I tend a couple of torn ears, work on hooves that are splitting, drain a nasty neck abscess, use a stethoscope on those with colds to make sure they aren't flirting with pneumonia or bronchitis. I like this part of my day because I get to work hands-on with the pigs and give some really good belly rubs. Today, they are glad to see me leave because I'm wet and smelly, and disturbing their naps.

3:30 p.m. to 5:30 p.m. Time to start feeding and watering again. The pigs are even grumpier because the weather has kept them cooped up all day. My forcing them out into the nasty weather instead of feeding them inside makes them even madder. The fact that I must sometimes shovel snow from their doors several times a day and cut paths for them to travel doesn't concern them.

Pigs don't do gratitude.

That used to bother Laura and me. Early on, we thought that a rescued pig should show their thanks with affection and sociability. Now, we truly understand that the gift is in the giving, which means letting each pig choose their own level of "people comfort." We love our cantankerous pigs as much as our cute, belly rub-loving ones . . . perhaps more so, because these are the pigs nobody wants. They, more than the others, need the safety and nurture that our sanctuary provides.

5:30 p.m. I am draining the last hose just as Laura arrives home from work. The cats are lined up at the front door, muttering and pacing. They know she will let them in where it's warm and fill their bowls. She will also feed Gus the Betta fish. Though it's dark, I take one last lap around the sanctuary with a flashlight to make sure there aren't any serious disagreements about sleeping arrangements. All our pigs are in their barns, except for a couple optimistically foraging for that last morsel of food, which they're sure can be found if only they are diligent enough.

I wait until they, too, are inside.

6 p.m. Over dinner, which, this evening, Laura has prepared, we discuss the pigs, our pending rescues, and other sanctuary business.

Construction on our new three-acre pasture was halted because of the weather. We have the posts, but the ground is still too frozen. Hopefully, in the next two weeks, we can begin. We already have about fifteen pigs awaiting its completion so they can be transported here. This pasture will comfortably hold about thirty pigs. With the fifteen scheduled to arrive over the next few months, some from as far away as New York, our population will grow to eighty-five. We will keep fifteen slots available for the most needy pigs, then stop when we reach one hundred, unless we can find more resources.

7 p.m. After dinner, Laura and I work on the books, pay some bills, and complete other administrative chores.

8 p.m. We have one more conversation—the one we dread. That's when we discuss which pigs we must turn away. Currently, we are refusing all "casual rescues," meaning pigs who aren't in any immediate danger. Their guardians simply want to get rid of them. In these cases, we attempt to problem-solve with them so the pigs can stay in their present homes. Failing that, we help place the pigs in good homes. So far, we've been successful, but finding alternatives for unwanted pet pigs is becoming more difficult. To date, we have not had to turn away a pig in a life-threatening situation, but we know that as we approach our hundred-pig limit, that day may not be far off.

This evening, we settle on two pigs but say no to six others.

Neither one of us likes playing God. While our minds tell us that we cannot save every pig, our hearts have not yet learned to accept that verdict . . . and probably never will learn. We are delighted to hear that a new sanctuary is being built in North Carolina that should relieve us of responsibility for taking in pigs from that state. We've also heard of another to be started in Maryland within the next six months.

9 p.m. Having dislodged several cats, I am dozing on the couch when the phone starts ringing. A board member wants to chat. A woman in the Midwest has been given a three-week-old piglet and wants to know how to treat him for diarrhea. Someone in the Northeast wants to know if pigs can be trained to do tricks. Someone from Ohio wants to chat about her pig's weight problem.

11 p.m. I nod through the news, then perk up for the weather report: more ice and snow for tomorrow. Temperatures in the teens, with windchill below zero, but rising into the forties on the weekend, assuring I'll be up to my ankles in muck. My final thought before sleep is a fervent prayer for the pigs living here, as well as for the millions of needy pigs around the world. I ask for the patience, health, and funds to do what little we can to ease suffering for as many as we can.

All in all, it has been a good day. There are no bad days here at the sanctuary. Any day that can be spent caring for the pigs is a good day. And tomorrow, in spite of the weather, will be yet another good day.

5

Big Earl and Me: A Bromance

IN JUNE OF 2001, MINI-PIGS received a call that would radically change our rescuing mission. A woman tearfully told one of our board members of a farm piglet whom she had rescued from a commercial farmer in her area. Because of health problems, including a stiff leg with no articulation and a prolapsed umbilical cord, this piglet had been discarded in a ditch where she would have starved to death. The woman had taken in the piglet, whom she had named Babe, inspired by the 1995 movie, and she had hand-raised her. At just under one year, all Babe's health problems had been resolved, except for the stiff leg. Now, however, she faced a dire threat. The woman's husband said he would butcher and eat Babe unless another home could be found for her.

Babe's time would be up on June 16—only two weeks away.

Laura and I had been turning away pigs due to limited space and funds. Our entire experience was with miniatures. We didn't know what to expect from a Yorkshire pig who might grow to be 800 to 1,200 pounds. Still, that death date had a powerful effect on our psyches. So did our knowledge that more than 2.3 million farmed pigs are slaughtered each week to feed the "pork habit" of Americans. Babe became symbolic of this holocaust-like number.

We agreed to take her.

All the experts had warned us that you can't mix farmed pigs and miniatures. They just don't get along. When Babe—a white Yorkshire with upright ears—arrived at the sanctuary, we didn't have a separate space for her. Since she was crippled, we decided to put her in a barn with the elderly potbellies, who weren't all that rambunctious.

The next morning, I felt some anxiety as I went to investigate. Babe was asleep in the hay. All the potbellies were snuggled up to her. Babe was a potbelly magnet!

From that moment on, Laura and I realized that potbellies and farmed pigs could live together without a problem. We also fell in love with pigs all over again. Babe was so dramatically different from our potbellies, and it wasn't just her size. She was so laid-back, whereas miniatures were drama queens, perhaps due to their little-guy, Napoleon complex.

It was our vet, Dr. Tom Massie Jr., who persuaded me to take in our next farmed pig as a companion for Babe. His name was Big Earl and he was massive, probably 1,200 pounds. He belonged to a family that raised horses. Though the family loved Big Earl, they couldn't give him a good home. The horses didn't like him, so Big Earl was lonely.

All of the animals in my life have been special. But if you are an animal lover, once or twice in your lifetime, you may come across an animal who rocks your world. A special bond develops between you and that animal that transcends species. For lack of a better explanation, I'll call it a meeting of souls. Big Earl was my first experience with this phenomenon. When the doors of his trailer opened and that big Yorkshire stepped out, it was like something cosmic happened. With his great bulk and huge, floppy ears, Big Earl seemed like a behemoth next to our other pigs. The rapport between us was instantaneous and powerful.

Big Earl tried to make friends with Babe as prescribed. Perhaps because Babe was disabled, she didn't want a pal who was 500 pounds bigger than she was. Babe would snap at Big Earl, and he was such a good-natured fellow that he'd run from her, screaming. Not yet satisfied, she'd hobble after him and bite him on the ass. Though Babe still let the potbellies hang around her, even sleep on her, she didn't want any part of a relationship with Big Earl.

I think that rejection bothered Earl, so he became my constant companion. He followed me everywhere, worse than if he were a puppy. May I say there is nothing more fun than crawling under a tractor to fix a broken hydraulic line while a big farmed pig unties your boot laces, sticks his head under the tractor to offer technical advice, and rearranges all your carefully-laid-out tools. With help like that, my one-hour chores could wind up taking four hours. No matter what I was doing, Big Earl would be either by my side or nearby. Going up the long driveway to check the mail? Big Earl would tag along. Repairing a fence? Big Earl would be at my shoulder scattering fence staples. At night, he would try to follow me into the house, succeeding on several occasions, only to be lured back outside with a handful of vanilla wafers.

When I was too busy to give Big Earl the attention he craved, he struck up a relationship with Scooter, a tiny, affectionate, and very vocal male potbelly someone grew tired of and booted out of the house. Scooter was white with black and brown spots and an itty-bitty corkscrew tail. Since he weighed no more than seventy-five pounds, it was comical to see Scooter and Big Earl eat out of the same dish.

The two of them first met when Scooter was young enough to be living in the basement of our house. One evening, while I was cleaning up down there, I heard this strange banging noise. It was Big Earl. He had leaned against the French doors with his heavy body until they gave way. After wandering inside, he decided he liked the warmth and was making a nest for himself. While Scooter and the other babies watched, entranced, I tugged, pushed, shoved—being outweighed four-to-one—trying persuade Big Earl to change his mind. Again, half a box of vanilla wafers convinced him it might be a good idea to leave.

When Scooter moved outdoors, he and Big Earl had the pleasure of each other's company once again. That happened after I had spent a day building a wrap-around porch and a handicap (read: "pig") ramp into the house. The next morning, as I was testing the ramp for weight, I heard loud squeals of protest from under my feet. I discovered Scooter wedged under the ramp, where he had apparently crawled for a nap then ended up having to spend the night. He was wet, cold, and very irate. His pudgy little body was stuck so firmly that I had no option but to dismantle much of the ramp to free him. As I was completing that rescue, I noticed muddy hoof prints all over my new wrap-around porch floor—undoubtedly Big Earl's. The rattan furniture had been overturned, and Laura's cushions—now muddied, with tufts sticking out—had been made into a nest. Apparently, Scooter and Big Earl had been camping out on the new front porch.

The next chapter in our farmed pig story began when we learned of a Hampshire named Leo. Leo had been a breeding boar at one of the popular museum farms set up to replicate particular time periods—in his case, as it so happened, the Revolutionary War era. Because Leo could no longer breed, the museum planned to send him to slaughter. However, the community had come to love this massive, gentle, big black Hampshire with white forelegs and shoulders. When I received a call asking if we would take him, I was by then smitten enough with farmed pigs to agree.

Leo, who had been confined to a pen for his entire life, could barely walk. He was also massively obese. At our sanctuary, where he had his freedom, he struck up a good relationship with Big Earl. After Leo had lost weight and become more mobile, I'd watch Earl and Leo head down the road to our stream, often with Scooter tagging along. Since Leo couldn't walk as fast as the other two, Earl and Scooter would get maybe a hundred yards ahead. Earl would turn and holler at Leo. Leo would amble along as best he could. Earl would then go all the way back to Leo to urge him forward. After a while, Leo's mobility improved to the point that he could saunter along with the others. Two 1,200-pound pigs bookending 75-pound Scooter—an unlikely trio of buddies.

At the stream, the big guys had begun fashioning a mud hole that was an architectural marvel. It had a deep end. It had a shallow end. It had a shady side. It had a sunny side. As always, I was intrigued by the personal attention that went into the creation of these spas. At about midday, I'd often fill my pockets with apples, then wander down to the mud hole to hang out with the guys for a while. By then, they were usually up on the bank, in the shade under a tree, sleeping. I would share my apples and spend the afternoon under that same shady tree. It was Big Earl and Leo, maybe Scooter, and me as an unlikely fourth. No words can describe the pleasure and contentment of dozing between two gigantic pigs on a warm, sleepy summer afternoon.

I think the large size of farmed pigs had something to do with their docile nature. Perhaps they had also been genetically engineered for that trait. If these marvelous creatures managed to somehow escape slaughter by the tender age of six to eight months, they had precious few options. I made up my mind to bring in more of them.

As the sanctuary's reputation as a haven for farmed pigs grew, we began to receive calls from animal-rescue groups all over the country. As soon as I heard the words, "Mr. Hoyle, I'm calling from the Liberation Group—" I would stop the caller right there. I'd tell them: "I don't want to hear any of the details. Just tell me your name is Joe Smith. Tell me you have a farmed pig you need me to take, then let's make the logistical arrangements to bring your pig here."

I didn't want to be told about some felony that might have been committed to free this pig. As much as I would have liked to springboard a public awareness campaign, I didn't want to endanger my sanctuary by knowingly taking in stolen "property." I guess you could call it willful ignorance. Politicians might label it "plausible deniability," meaning a denial of responsibility for the actions of others because of a lack of evidence confirming one's participation.

Big Rosie came to us after falling off a transport truck on her way to slaughter. As a farmed pig, having been fattened for consumption, she was so obese she could barely walk. Her hooves were extremely long, and she was covered with road rash from her highway accident.

With good nutrition and a chance to socialize and exercise, Big Rosie made an amazing recovery. This included the loss of 150 pounds despite her love of peanut butter and jelly sandwiches, occasionally allowed as a treat. It was a delight to see Rosie loping around the pasture with little regard for hooves and legs that would have been deformed from carrying all her former weight.

Midnight, a potbelly, came to us from South Florida via a "piggy railroad" of caring people who transported him up the coast. Because he had seizures, we quarantined him in our baby pen for over six months for evaluation. Since that pen had been reseeded, it was now filled knee-high with clover, alfalfa, fescue, orchard grass, and timothy. It had two other occupants—our rabbits, Mr. Hatter and Mr. Bunny Man. Though they couldn't quite figure out why this strange-looking creature was joining them, all three soon became friends.

Midnight was small—only about sixty pounds—and timid even for a potbelly. Though he was afraid of all our other pigs, once he was out of his quarantine pen, he formed a remarkable friendship with sweet and cuddly Big Rosie. They were always together. Rosie even tried to get into Midnight's dogloo with him. She could only just manage to push her head and shoulders in, with her butt hanging out. Like Earl and Leo and Scooter, these two were a comical sight, one so big and the other so tiny. Midnight only had a couple of seizures after coming to the sanctuary, allowing us to wean him off all his medications.

By the time we were celebrating our third anniversary in the foothills of the Blue Ridge Mountains, we were home to approximately one hundred fifty miniatures, four farmed pigs, eight rescue dogs, forty-five rescue cats, one pygmy goat, two rabbits, and our Betta fish Gus.

We had added four pastures, a number of quarantine pens, and eight more small barns to house new arrivals. Each day during the winter months, we distributed approximately two hundred pounds of pig feed, along with eight to ten pounds of dog food, twelve pounds of cat food, and four bales of hay. Twelve pigs were on daily medication for a variety of chronic problems ranging from arthritis to stomach ulcers, the result of poor treatment by former "owners."

While welcoming farmed pigs, we continued our commitment to potbellies. Larry, Curly, and Moe were three ex-boars who arrived together. They were accompanied by Honey, a dog someone had tossed out of a moving car. These three scruffy little boys were half wild and very full of themselves. They immediately bonded with Noel's babies, now out in a pasture, creating a little street gang of roving thugs, always in action and frequently into mischief.

Next came six pigs from a nine-member gang that had been running wild on Catoctin Mountain in Maryland. Unfortunately, two of the boars had been killed by locals before a rescue crew could capture the other seven. This was a team effort, including shared expenses. We retrieved the females, whom we named Ellie Mae and Daisy Mae. The five big boars temporarily went to a sanctuary in Joppa, Maryland, to be neutered, vaccinated, and dewormed and have their tusks trimmed. After quarantine, four of the five were delivered to Mini-Pigs. Though Tommy, Wynken, Blynken, and Nod were tough-looking bruisers, with scars all over their bodies, they soon turned into amiable big guys, content with life in their four acres of woods.

Both the sows, Ellie Mae and Daisy Mae, were fat-blind from poor diets. A week after their arrival, Ellie Mae escaped from quarantine. We found her in the deepest recess of the woods, frantically making a nest to farrow. We provided her with several bales of straw, then erected a tarp over her head and a fence around her to give her privacy.

Early the next morning, Ellie Mae presented us with eight very premature piglets, two of whom were stillborn. We watched in awe as she took each of her dead babies in her mouth, then walked with them to the farthest corner of her pen, where she dug a grave with her snout. Then, after burying her babies, she lay down on their graves, making soft grunting sounds for a few minutes before returning to nurse her surviving piglets.

When two more of Ellie Mae's piglets died, she buried each in the same way, a deeply moving portrait of maternal grief that transcended species.

Afterwards, we enlarged Ellie Mae's pen, where she raised her four surviving, healthy, wild little piglets. When weaned, the two boars were neutered, and all four were moved to the Baby Pen, either to be adopted or to join a herd when able to hold their own.

Ellie Mae's piglets were among the twenty-two born that year at the sanctuary. Thankfully, our record of no accidental breeding remained intact. All came from females who were pregnant when they arrived.

Precious, a tiny four-month-old female, wandered into a subdivision after being abandoned by her guardian. Some compassionate folk contacted us after finding her rooting in neighborhood gardens. Precious was a "wild child" who had never been socialized, so we worked with her to help her enjoy being around people. Like many of our new arrivals, she wanted to be friendly, but learning to trust humans after being callously treated took time.

Laverne and Shirley were two young sisters who had also been abandoned. They, too, had been rescued by kind folk who couldn't keep them because of zoning laws. Both girls were very friendly and cute.

The sad truth about life at a sanctuary is that space often becomes available for newcomers because old-timers and the sick or injured die.

Leo, one of our beloved Hampshire farmed pigs, passed suddenly in April of 2003 after only two years of retirement. Because this happened on a Saturday, his necropsy was done at the sanctuary, and we buried him in a special spot back in the woods where he loved to roam. Leo received a front-page obituary in the Morristown, NJ, newspaper in recognition of his popularity at the Revolutionary War-era museum farm where he had lived most of his life.

Earl grieved for his pal for weeks.

A year later, I was hit with an even more personal loss.

As I'm a proud Irishman, St. Patrick's Day is usually a time for me to celebrate my heritage, but not on March 17, 2004.

In late February, we had had a really bad cold snap. All our pigs, especially the big farmed pigs, stopped traveling to water because they didn't like the treacherous ice underfoot. Now, a healthy pig will drink between 10 and 15 percent of their body weight every day. Though

Laura and I spent each waking minute hauling water, it was physically impossible to supply our large number of pigs with all they needed. Compounding the problem was the fact that pigs do not like really cold water. Even when it's nearby in buckets, most won't drink enough.

So, despite having water available, Big Earl became constipated, leading to an intestinal blockage. For two weeks, Dr. Tom and I tried to get the blockage to move by various means. Finally, Tom advised, "You should take Big Earl in for surgery."

I drove Big Earl to Virginia Tech College of Veterinary Medicine, where they performed the operation. I was in surgery with him for the four hours it took to resolve the blockage, and then I went with him to recovery. I was there, holding Big Earl's huge head in my lap, when he took his last breath.

This was one of those medical cases in which the operation is a success but the patient dies. The anesthesiologist had not bothered to intubate Earl for his lengthy surgery. While on the operating table, Earl had become hypoxic, meaning that he couldn't adequately ventilate himself, which deprived his tissues and vital organs of sufficient oxygen. Post-operative pulmonary edema set in, killing him.

Virginia Tech asked to do a necropsy on Earl, to which I agreed. But first, I wanted a memento. Earl had very long tusks. After requesting a cutting tool, I carried away those tusks, leaving the rest of Big Earl with Virginia Tech.

It was a six-hour drive back home. I didn't get out of the college until eight o'clock in the evening, and I had to be at work at six the next morning. I drove all night through a snowstorm, pulling the empty trailer. Back at the sanctuary, I changed clothes, then headed off to the fire station.

Losing Big Earl was like losing a brother who was also my best friend. Quite frankly, if I could have gotten my hands on the anesthesiologist at Virginia Tech who had failed to do his job, I might have acted in a way that would have landed me in jail. To me, he was guilty of gross incompetence. Like too many other veterinarians, he thought that a pig was "just a pig," so not worth his best professional effort.

Big Earl and Me

Written on the passing of Big Earl,
March 17, 2004

When the trailer pulled up and they dropped the gate
I knew his arrival was more than fate
One look in his eyes and I could plainly tell
His life with us would go very well
A special bond there soon would be
Between this pig, Big Earl, and me

He came as a companion for the big pig, Babe
But she don't care for this big white knave
She chases him and runs him ragged
And bites his butt till his nerves are jagged
He's a gentle old man as all can see
We're the best of friends, Big Earl and me

He's long and tall and very sweet
He's a lot of things, but not petite
He's as laid-back as a pig can be
Like me, he's clumsy as can be
Can't neither of us climb a tree
We're two of a kind, Big Earl and me

He helps me with my chores each day
Most of the time he's in my way
I'd chase him off but I'd rather he stay
While across my tools he likes to lay
He helps me work till time to feed
We're quite a team, Big Earl and me

Some days when the weather is warm and hazy
Earl and I get kinda lazy
We find a tree that's big and shady
We lay down together in its lee
And take a snooze, Big Earl and me

He lays his head upon my lap
And both of us take a little nap
We talk and talk 'bout this and that
I look forward to our daily chat
He's a pretty smart pig I've come to see
We have a good time, Big Earl and me

We talk at length of all his kin
Butchered by men on just a whim
To see his brothers upon a plate
You'd think would give him cause to hate
But pigs can't hate, I've come to see
We've learned a lot, Big Earl and me

There's a lot we just don't understand
Why man must kill to feel so grand
And always have the upper hand
He's taught me lots 'bout humility
And different ways I've come to see
Thanks to the love 'tween Earl and me

When our race is run and we meet out fate
We'll both show up at the Pearly Gate
I'll bet Saint Peter will make us wait
While he asks the Lord to look and see
If there are two heavenly spaces free
One for Big Earl and one for me

And if the answer is, "Y'all come on in"
We'll both enter heaven free from sin
We'll nose around for a while to see
If we can find a celestial tree
We'll kick back all trouble-free
And enjoy heaven together . . . Big Earl and me

But if the word comes down, "No pigs in here"
We'll tell Saint Peter, "Stick it in your ear"
We'll pack our gear and keep on lookin'
Till we find a place where pigs are rootin'
It won't be heaven, don't you see
Unless there's room for Big Earl and me

I could pen many verses longer and cuter
But Earl's lost interest and gone off with Scooter
To do pig things far more entertaining
And left me here with rhymes waning
So I'll leave you now and go and see
If I can make it three . . . Scooter, Big Earl and me.

6

Just One More Pig

LAURA AND I REALIZED EARLY on that we were good at providing animal care and that we truly loved working with the pigs. My years in the military and firefighting had prepared me to accept crises as a normal part of daily existence. We didn't mind farm chores, and we were both adept at running and fixing machinery. Laura's training as an emergency medical technician and mine as a paramedic gave us an edge in dealing with medical problems. Laura had developed administrative skills from her hospital work, and I had a master's degree in organizational development and human resources. Yet nothing in our work or educational experience taught us how to generate the funds to supply, say, 250 pounds of feed a day to support 160 pigs, many of whom arrived sick and emaciated and in need of medical attention.

Laura and I continued to absorb all the costs of running the sanctuary. We used donor funds only for the direct care of the pigs, which meant feeding them, paying their very substantial veterinary bills, providing fencing for the pastures, and building their barns. I'm lousy at asking for money. I just don't have the skill set, and neither does Laura. We kept at it and somehow managed to end each year in the black—or just slightly in the red. Though this translated to a frugal lifestyle for us, I believe these were the happiest and most fulfilling years of our lives.

In hopes of raising ready cash, we held weekend visitors' days during Virginia's months of good weather.

I dreaded them.

Perhaps eighty boisterous city folk, who knew nothing about farmed animals or farm equipment, would descend upon us. Given our solitary life at the sanctuary, I wasn't used to interacting with that many people under the best of circumstances. What I saw, climbing out of cars and walking towards us, was an alien invasion. Often, it included young children or, as I generally referred to them, "tiny, highly mobile germ factories" with very busy hands.

Laura and I, and one or two board members, would take our guests on guided tours from pasture to pasture, introducing them to various pigs and telling pig stories. To our guests' credit, most wanted to pitch in with chores. Trying to find something useful for them to do might end up with a poop-raking contest. Our pigs loved these tours because pig-cooperation always meant tasty treats.

It was impossible for us to keep an eye on eighty guests, no matter how diligent we were. Some would do really clueless things: One mother was determined to take her three little kids into a pasture for a picnic of peanut butter sandwiches. She might as well have covered her kids with blood and thrown them to a pack of great white sharks. Pigs love to eat—*oh, how they love to eat*—and nuts are their favorites. Led by their unerring snouts, they would have trampled those kids and devoured those crunchy sandwiches, not because they're aggressive but because they're genetically driven by their ferocious appetites.

As we finally waved goodbye to our sun-burned, bug-bitten guests, then watched the last set of tail lights disappear out the gate and over the hill, I would be ready for a couple of good shots of Irish whiskey. And sometimes, after these grueling weekends, the donations barely covered the cost of holding the event.

We were much more successful with our educational goal, which was two-pronged: to raise awareness of the plight of animals in the factory-farm system, and to promote a plant-based diet. Most people who interacted one on one with our pigs—especially our farmed pigs—left the sanctuary touched by that experience. We were often amazed to receive an email or a phone call after maybe five years from someone who had been inspired by a specific incident that hadn't stood out to us. Perhaps they had gone on to join an animal rights group, or had stopped eating pork on their way to becoming vegetarian and eventually vegan.

As part of our outreach, we also developed a program for student interns to spend four months with us from spring to fall. Some only lasted a month, but several others went on to become veterinarians. It was wonderful to see these kids who might have never had an opportunity to work with animals blossom and grow. They, too, could look into a pig's eyes and watch their body movement, and understand what that animal was communicating, and feel an unconditional love and trust. Others came to us as part of their veterinarian training. This internship program was something that we, as a sanctuary, thought we should do. The reality was that it often took up our time and took away from what Laura and I were able to accomplish while still functioning as a hospital administrator and a firefighter.

In the fall and in the spring, the sanctuary held a volunteer work weekend to prepare for winter and to recover from the property damage created by winter storms. Like our weekend visitors, most volunteers were city folk from Charlottesville, Richmond, and suburban Washington, DC, with few usable farm skills, so, again, we assigned them to tasks such as scooping and moving manure. Over time, we did cultivate a group of about ten regular volunteers who didn't have to be constantly supervised. We would assign them to groups of other volunteers to repair a barn roof or to fix fences. The sanctuary supplied a vegan lunch and pots of tea, coffee, and fruit juice. Some of our visitors camped on our property and brought musical instruments for evening jam sessions. Those weekends often turned out well for us.

Originally, Laura and I fenced only our pastures; however, during our eight years in Culpeper, we managed to enclose our whole seventeen acres with a high-quality, sturdy perimeter fence so that our farmed pigs could roam the entire property at will. We built our own barns. We did everything we could to make the operation of the sanctuary more efficient in the limited time we could take from our professional lives. Because we didn't have enough land to grow hay, I was buying it from a farmer down the road. That meant having to transport and store it. We also did a lot of hauling of feed, until I was able to persuade a local co-op to mix and deliver a feed that I designed. Then, all I had to do was unload the truck every week and measure it out for each pasture. Though these innovations helped with the sanctuary's daily chores, many times I'd come home at 9 a.m. after a twenty-four–hour shift, with no sleep thanks to back-to-back emergency calls. I'd grab a couple of cups of coffee, then work in the pastures and barns until bedtime.

A major 2001 project designed by several volunteers ran waterlines from our well pump into each of our pastures. A board member and her husband obtained all the necessary materials, including piping and yard hydrants. Another volunteer provided a large trench-digging machine to bury the pipe. As we laid the waterlines, board members and volunteers filled in more than one thousand feet of four-foot–deep crisscrossing trenches so our pigs wouldn't fall into them. All this was so successful that the following year, we ran electricity to each pasture. That meant every pasture had its own hydrant for water and an electric post where I could plug in stock tank heaters to keep water from freezing. It was a huge project and a marvelous display of love, which dramatically increased our efficiency.

We also had volunteers who helped with pigs who had trouble adapting to sanctuary life. It was because we had taken in a seven-year-old, nearly blind, overweight potbelly that we came to know the guy who supplied the trenching machine. Hazel was slowly enticed from her world of confusion by the reassuring voice of a volunteer who read to her every day. Hours of reading also helped Ella, who was labeled a "mean girl" because of a traumatic background that made her nip at both humans and other pigs, understand that whatever had made her so unhappy in her old environment hadn't followed her here. With her guard slowly melting, she became a chill, happy-go-lucky girl who was able to enjoy both human and pig companionship. In these therapeutic sessions with frightened pigs, some of our volunteers shared stories they had enjoyed as kids. Others read classical literature, or whatever they happened to be reading for themselves. The words didn't matter to the pigs, just the peaceful human presence and the sound of a soothing voice.

To raise money for Mini-Pigs, I wrote a regular column for *The Herd*, a national potbelly pig newsletter, which proved successful in bringing in more donors. With Irish blood on both sides of my family, the gift of the blarney was strong in me. We Irish don't know how to make a long story short. If you ask us the time, we might wind up telling you how to build a clock. This exposure led to a few public-speaking engagements, along with interviews on TV stations in the Washington, DC metro area. A five-minute spot on the local news about our unique little sanctuary might bring in four or five new donors . . . along with ten to fifteen calls from people wanting to dump their pigs on us. Definitely a mixed blessing.

For our first vegetarian festival, Laura and I put up a modest booth, to which several board members contributed time and money. We designed brochures and printed copies of our newsletter. We had a professional banner and a large pictorial display of photos of our pigs and our sanctuary. The event lasted ten hours, and we talked to several hundred people, all of whom seemed excited about what our sanctuary was doing. The take for the day was $68.15, not enough to cover our travel expenses. Still, we persisted, with Laura usually being the one attending on behalf of Mini-Pigs, perhaps accompanied by one or two board members or volunteers. These efforts, in their aggregate, were significant, since we existed mostly on a steady flow of $5, $10, and $25 gifts, with the occasional surprise donation of several hundred dollars.

Grants for Mini-Pigs were elusive. Of the thousands of philanthropic organizations, only a small percentage support animal rescue. An infinitesimally smaller percentage fund farmed-animal rescue, and most of those state on their websites that they are no longer accepting applications for funding. Our attempt to engage corporate sponsors was haunted by the unfair stereotype of pigs as dirty, slovenly animals, unsuitable for corporate images. As unrepentant optimists, we kept firing off hope-filled applications, fueled by that little/big word—"IF." *IF we had enough money, we could finish the new pasture. IF we had four more barns* . . . . IF we put all these "IFs" into one sentence, it would read: *IF we had more money, fewer pigs would have to die.* Once in a great while, we were rewarded with a small grant for a new barn or to help with a high vet bill, but we were mostly ignored or turned down.

In seeking donations—begging for them—Mini-Pigs was in a very competitive field. I was amazed at the dozens of professionally prepared mail-outs that Laura and I received each week from every conceivable animal welfare or animal rights organization. If Mini-Pigs had had the funds it cost to produce even one of these, we could have rescued a hundred needy pigs.

It also saddened us to discover how many people who professed to love animals dug deep into their pockets to fund a rescue for dogs or cats but scoffed at the idea of supporting a sanctuary that rescues pigs.

If you operate a mom-and-pop sanctuary, you perpetually live under the Sword of Damocles, and that sword is suspended on a very thin thread. Many of our loyal, longtime supporters were working folk without much money to spare. They donated because they believed in what we were doing, but when hardship struck, charitable giving was one of the first items to be dropped from family expenses already stretched thin.

Our funding was also dramatically impacted by cataclysmic events, such as 9/11. After that terrorist attack in 2001, donations plummeted to nearly zero, not only for our sanctuary but for all animal groups, with money being shunted off to disaster- and victim-relief organizations. Recessions (and more recently, pandemics) have a more prolonged effect.

Unless sanctuary owners are independently wealthy, and few of us are, failure to raise funds leaves us with two options: to watch our animals starve to death, or to euthanize them. This is not mere speculation. It has happened to a number of miniature-pig and small farmed-animal sanctuaries in this country. On hearing of these tragedies, my first reaction is horror, followed by a gut-wrenching sadness. Next comes anger, especially if I learn the sanctuary owners continued to "rescue" while their herds and their finances were deteriorating. Finally, I experience frustration, helplessness, and futility as I wrack my brain for ways I can help. Often, we have taken in pigs from a failed sanctuary; however, the lack of space and/or money sometimes didn't allow this option, regardless of how heart-wrenching the circumstances.

Because of these tragic examples, Laura and I always paid strict attention to the advice of our board. About a month before each year's end, we presented them with an accounting sheet for reviewing our books, which were also left open for our donors. Together with our board, we developed three-, five-, and ten-year plans, to be reviewed annually to see if our goals remained valid or if they needed to be adjusted.

The board gave us free rein to spend up to a certain amount, but beyond that, we needed majority approval. Though all our board members supported animal rescue in principle, they did not have Laura's and my overwhelming passion. We counted on them to keep us grounded in reality because if you let a novice sanctuary director go unchecked, they'll always find room for one more pig and can "just one more pig" their way into bankruptcy.

Laura and I were proud when Mini-Pigs was accepted as a member of The Association of Sanctuaries (TAOS), a prestigious accreditation organization for professional sanctuaries. Our final on-site inspection was conducted by a veterinarian and an animal-welfare investigator, who were favorably impressed with what they observed. Their recommendation was the last step in a two-year process, recognizing Mini-Pigs as a professionally run sanctuary with extremely high standards for the care of our animals.

As I previously explained, early in the life of Mini-Pigs, we began an adoption program. Since our potbellies had been bred as companion animals, we felt it made sense for us to rehome them whenever possible. Our rules were stringent. We would only provide a single pig to someone who already had one pig and was looking for a companion. Otherwise, they had to adopt a pair. Our trained volunteers vetted each prospective family with a list of requirements, including the necessity of having a local vet who would accept their pig as a patient. Each adopter was required to sign an ironclad legal contract, giving us the right to take back their pig in case of abuse or neglect. If the adopter could not keep the pig for any reason, that person was required to contact us within seventy-two hours. It then became our legal duty to retrieve the pig in less than five days.

One of our happiest adoptions involved little Scooter, who was so lonely after Leo and Big Earl died. We adopted him and a tiny female named Emma to a really nice older gentleman who had just lost his ten-year-old potbelly to cancer. Scooter and Emma became much loved and pampered little pigs under their new guardian's doting attention.

The more our sanctuary grew, and the more pigs we had for potential adoption, the harder it became to run our program the way it needed to be run. If we had, say, twenty pigs in homes, that would mean about forty follow-up visits. Given Laura's and my jobs, along with the time it took to run the sanctuary, we didn't have hours enough for those critical follow-ups. Out of approximately one hundred adoptions, we retrieved maybe eight to ten pigs from unsuitable circumstances. Those usually involved someone who had begged me for a pig, then didn't want to admit that the adoption wasn't working. Perhaps there was a behavioral problem, or perhaps the pig had become sick, and the adopter had lied about having a competent vet.

We had one pig who died after his adoption. When Roscoe quit eating, his adopters consulted their vet but didn't take him in for a check-up. By the time they called us, the problem had been ongoing for a month. Roscoe had become constipated, and then blocked, because he hadn't been receiving sufficient fresh water or adequate veterinary care. I reclaimed Roscoe, then rushed him in for surgery to remove the blockage. After several thousand dollars, we lost poor Roscoe anyway.

It was the loss of that one pig that ended the program for me. If you have, say, fifty pigs adopted out and you aren't on top of every case, statistically, two pigs are likely to fall through the cracks. Now, some sanctuary owners might say that's a pretty good average. You've found homes for forty-eight pigs out of fifty, who might have ended up dying if they hadn't been rescued and adopted. To me, that wasn't an acceptable percentage. I wasn't willing to lose those two statistical pigs, or even the one we did lose. From that point on, each pig we welcomed into our sanctuary came with a lifetime commitment. What had seemed like a reasonable plan turned out not to be satisfactory. At Mini-Pigs, we wrote our rulebook in pencil, for which we kept on hand a bountiful supply of erasers.

When Laura and I had adopted Patti Murphy, a miniature pig's normal lifespan was considered to be eight to ten years. At Mini-Pigs, with a nutritious diet and expert veterinarian care, we increased that lifespan to fourteen to eighteen years, even up to twenty to twenty-two years in exceptional cases.

Though Patti and Stinky had drifted apart over the years, we would sometimes find them sleeping together under a tree, rekindling the bond they had forged in earlier years. As our first rescue, Stinky held a special place in our hearts, so I was both surprised and heartbroken, during a morning feed in 2001, to find him lying dead in a pasture. Stinky was fourteen years old and had never had a sick day in his life. His death caused Laura and me to reflect back to the time when our "herd" had consisted of seven members—two rambunctious pigs, three dogs, and two very overwhelmed and perpetually confused human adults.

A necropsy performed by the state veterinarian's office failed to yield any cause of death.

When Patti realized Stinky wasn't around anymore, she wandered about, as if looking for him, and moped. Finally, after about ten days, she reverted to being her normal self, but she never formed a deep bond with any other pig for the remainder of her life.

7

Prayers, and a Surprise for Patti

I SIT AT MY COMPUTER with my coffee early this spring morning in 2003, watching the sun come up on the sanctuary. Our office (the spare bedroom) overlooks two of the pastures, where I will soon see our pigs start to awaken. It is a serene time of day. I let Samantha, our new Labrador puppy, out at 5 a.m. Now she is busy investigating every smell and leaf while trying to get the other dogs to play with her. How I wish I were blessed with puppy energy!

I never tire of watching the sanctuary come alive in the morning. The pastures are greening; the trees are budding out. Our wild turkeys are breakfasting on the uneaten morsels of corn and other grains they pick up under the feed troughs. Our resident flock of crows will soon make its morning flyover.

Usually, we can count on a beautiful spring, but this past winter was unforgiving, with spring a long time coming, making this morning all the more precious. Rain. It rained for a couple of months, all through early spring. On checking my calendar, I found that we hadn't had four consecutive days without precipitation since last October.

I need these mornings. They are a tonic for my soul. The days are spent working on the computer, answering emails, talking on the phone, greeting the occasional visitor, and, of course, meeting the constant demands of the sanctuary and the animals. After dark, long after the animals have been fed and watered and have fallen asleep in their barns, the administrative work of the sanctuary continues until both Laura and I are too tired to think straight.

But the mornings . . . they allow me to reflect without interruption.

In the Big Pig Herd, the six Ossabaw Island teenagers are always the first up. I see them now.

In the Little Pig Herd, it's Ellie Mae's offspring, Sally-Sue and Pig-Pig, who are the early risers.

As each animal emerges from their barn to greet the new day, I think back to the circumstances that brought them to us. Many of the stories are unrelentingly cruel, and each is unique. But all have one thing in common: the animals found peace, safety, and love here with us. Nobody else in the world wanted them, but somehow they managed to find their way here, often days or even hours before they were to be killed.

I can now hear the piglets in our Baby Pen stirring. Our babies have found the alfalfa I snuck in for them last night after dark. Now they are busy rooting around in it, spreading it over their pen, and munching contentedly on the green leaves.

Our older pigs like Elvis and Li'l Richard will be slower to rise, with the last ones being our crippled and arthritic pigs like Babe and Big Rosie. They always wait for the day to warm up before hobbling out for a morning drink of water.

When our free-roaming pigs—Patti Murphy and Snigglers—come wandering into the front yard, it will be time for me to shut down my computer to begin feeding. All our animals know exactly when it's time. If I'm late, they let me know it by squalling as if they're being murdered.

My thoughts invariably drift to the pigs we've had to turn down in the past weeks. What has happened to them? Why weren't they as lucky as the pigs who will soon be dotting our pastures? Could we have taken in "just a few more?" Did our refusal condemn a pig to death or a life of continued abuse by humans who no longer want to be bothered with them?

Tragically, we continue to read about miniature pigs, abandoned by their heartless "owners," being sold at commercial slaughter auctions.

Since they can't be used as food for humans, they are purchased for pennies by pet-food suppliers or by unscrupulous petting zoo owners to feed their lions, tigers, and other big cats. The numerous newspaper ads offering potbelly pigs "free to a good home" chill my blood as I know how those stories usually end.

I can't allow myself to dwell on these harsh realities. Instead, I must focus on the heartbreaking calls for rescue that keep coming in to the sanctuary, changing any "typical" day into an untypical day. Two days ago, a contact in Maryland phoned with an all-too-familiar story. A young boar named Phred was to be killed because his guardians no longer wanted him. Laura and I had been turning away a number of pigs each week, but a space had unexpectedly opened up and the "just one more" mentality kicked in.

Phred turned out to be a cute, very personable little fellow, probably three or four years old, with a good deal of "boar personality" and a love of human contact. He immediately "woofed" his way into our hearts with his foaming at the mouth, his little Michael Jackson moonwalk, and his habit of dropping instantly for belly rubs when touched.

Who could be so callous as to want to kill such an engaging little guy? Had trusting, unsuspecting Phred been mere days away from being tossed, dead or alive, into a cage of hungry tigers?

Our vet, Dr. Tom Massie Jr., had a spot open and arrived soon after we called him to neuter Phred. Like most sanctuaries, we are able to use only injectable anesthesia for routine medical procedures. After having successfully completed over fifty of these operations, we have become proficient at using Telazol.

Ah, but complacency will get you into trouble every time.

After estimating Phred's weight, we gave him the appropriate dose of the sedative, then lugged his inert little boar body to the operating table, which was the wooden platform on the back of my tractor. We secured Phred in the neutering position, scrubbed him and us, then proceeded with the surgery. When Phred gave some indication that he was not fully out, we administered a little more Telazol, assuming we had underestimated his weight.

Phred's testicles were where they were supposed to be, and the neutering went without a hitch. We then vaccinated Phred, dewormed him, and trimmed his tusks.

Halfway through that last procedure, Phred quit breathing, and his heart rate slowed to virtually nothing. Tom took off on the run to his truck for medical supplies. I positioned Phred so his airway was open, clamped my hand over his mouth, and began mouth-to-snout breathing. When I heard him exhale, I removed my mouth from his snout. My free hand on his chest could feel a good rise. Estimating where Phred's heart should be, I began compressing on his sternum using a modified human CPR formula of two breaths and five compressions.

Tom arrived with the Atropine and Epinephrine and injected these into Phred's tongue veins. Phred's heart picked up speed almost immediately. As I continued breathing for Phred, he fairly quickly went from a nasty shade of blue to one resembling pink, lining up with that old EMS saying for humans: "Pink is good, blue is bad."

Finally, Phred took a breath on his own . . . then another. After a couple more minutes, he was chugging along all by himself with only intermittent stoppages. Half an hour after that, his heart rate was up to around 120 beats a minute, and he was breathing about twelve times a minute, with only occasional coaxing.

When Phred was stable enough to be returned to his pen, I stayed by his side. About an hour later, Phred was up drinking water and acting as if nothing unusual had happened.

I survived with only a bad case of boar breath and post-emergency jitters.

Rescues like Phred's are especially precious, given the previous winter's threat to the survival of our sanctuary. It was our worst since 1998, when Mini-Pigs was established. First came icy rains that stayed with us, allowing no break—a steady, often torrential drenching that turned the sanctuary into a giant mud pit. Barns could not be cleaned, and the daily feeding and watering took many more hours as we tried to negotiate through the deep mud. Though we trucked in more than two hundred tons of gravel, most of it was sucked into the mud as quickly as we spread it. In many places, the mud was so deep that our tractor became mired. The weather forecasters provided little hope that this pattern would break anytime soon.

Operation of the sanctuary had to be pared back to the basics as we tried to keep our heads above water (literally). I readily admit to feeding the pigs only once on days when the weather was so absolutely horrible. No apologies. It's not necessarily a kindness to drag them out of their warm, dry barns in driving snow or rain to eat soggy pellets. Watering them, on the other hand, is a chore on which we must never scrimp. Even in the most inclement weather, our pigs need fresh water at least twice—frequently three or four times—each day. Tending to our special-needs pigs is another chore that can't be short-circuited. Many require medications or special diets to survive and must be checked several times a day.

The most disruptive of the snow and ice started after Christmas. All in all, it was a miserable January and a worse February. A series of small but persistent storms rolled through, each dumping enough snow to make things tough, but still, we were coping. Then, something even worse happened. Laura became ill, with a persistent fever and stomach pains. Several doctors could not discover what was wrong with her. Because they thought she might have a urinary tract infection, they treated her with antibiotics. After improving for a while, she would relapse, such that her condition became chronic. Meanwhile, we were in the middle of a cold snap that seemed like it would go on forever, with layer upon layer of snow and ice.

It was brutal.

With the very worst to come.

While I was working one of my twenty-four–hour firefighting shifts in Fairfax, Laura called to tell me that she was in excruciating pain and needed to go to emergency. On receiving permission to leave the firehouse, I drove home, put Laura in the truck, and took her to our local hospital. At about 10 o'clock that night, Laura was diagnosed with a perforated appendix. It had been leaking for quite some time, so she was going septic. Since Culpeper's hospital was small and rural, we considered transporting her to Fairfax; however, it was decided that she was too ill to make the trip. Her surgery to remove her ruptured appendix had to take place that night in Culpeper.

I was familiar enough with this emergency to know that Laura might die. Having been in the Marine Corps and the fire service, and running our sanctuary, I'd seen so much of death that I'd come to the fatalistic conclusion that it was random, with little rhyme or reason. I'd witnessed auto crashes in which everybody should have been killed, only to see the occupants emerge with only minor cuts and bruises. I'd been called to what appeared to be an innocuous fender-bender with two people dead inside. Through the years, I'd learned to accept the inevitability of death. When it's our time, we can't do anything about it. It's out of our hands.

Perhaps it's odd that someone like myself, who was raised Roman Catholic but who'd rarely entered a church in decades, should talk to God a dozen times a day. And not just to God. I also have my list of Catholic saints to implore, like St. Francis for the animals.

I prayed during crises at the sanctuary when I couldn't see any way ahead.

I prayed when our animals were sick, and later over their bodies after they had died.

Laura used to laugh at the thought of me as a kid in a black and white cassock, trotting after priests chanting in Latin, but now I was praying hard and long for Laura. I knew she might die. Yet, somehow, I didn't feel that God was going to take her from me. I just didn't believe God would do that.

As it turned out, God didn't.

Laura survived the surgery. During the three days she was in the hospital, I cared for the animals then rushed to her bedside. When released, Laura was still on IV fluids, my paramedics training having factored into that decision. I took a couple of days off work to bring her home and to settle her for her "house arrest."

This coincided with one of Virginia's freak blizzards, dumping four feet of snow on the sanctuary. It blocked all our pasture gates and drifted over the doors of the pig barns. Several of these huts collapsed under the weight, trapping pigs inside. It took a full day of frantic shoveling to free them all, and then the snow was so deep that they had nowhere to go. I put our four-wheel drive tractor to work, plowing and breaking trails for them so I could distribute water. Even our tractor had trouble negotiating the pastures. Feeding started at first light, then continued after dark. Where possible, I fed in the barns. Otherwise, I dragged the feed troughs and buckets as close to the barns as I could.

With the bitterly cold weather, Midnight and two other pigs came down with pneumonia. Syringes of antibiotics had to be prepared several times a day. I carried them, warmed under my jacket, out to the barns, then crawled over the healthy pigs to reach the sick ones. Many late nights I spent sitting up with them. Thankfully, all recovered.

In the very worst of the bad weather, one of our farmed females suffered a prolapsed rectum. Since the road into the sanctuary was impassable for any vehicle, I trucked Tom and several armloads of medical equipment on the back of the tractor down the long driveway and into the back pasture. We performed surgery in the barn by flashlight and a lantern. Our girl was back on her feet and recovering by next morning.

Our sanctuary's feed supplies were running low, and the co-op could not deliver any further than the main road. Since feeding, watering, and caring for sick pigs were consuming every minute of my day, I had no time to clear our road. When I tried hauling feed on rough terrain from the main road on the back of our tractor, its weight caused the tractor's carry-all to collapse and break. The chances of my providing for our pigs looked grim.

Jim Greene—a neighbor, a good friend, and a pig lover from Charlottesville—heard of our plight. Amazingly, he dropped everything to bring his small four-wheel drive tractor to help with clearing the sanctuary road, along with other chores. He was a life-saver.

A second major storm, on the heels of the first, dumped another eighteen inches or so of fresh snow. Several old fence posts split due to the weight and had to be replaced in the dark in the middle of the blizzard. The sanctuary suffered a great deal of damage while also incurring tremendously high feed bills, vet bills, and equipment repair bills. Then, I tore a back muscle lifting a patient on a firefighting shift. That required several weeks of injury leave. While Laura hobbled around recovering from her surgery, I hobbled around nursing my bad back. Feeding and caring for the animals were a torturous and slow process. Incredibly, after only two weeks of recovery, Laura was out working with me, side by side, with feed buckets and bales of straw—*what a gutsy broad!*

It was touch-and-go, but we were still young enough and strong enough to survive. I also believe that period of life-or-death challenge drew Laura and me closer together. We needed each other on a moment-to-moment basis. We were there for each other, and together we saved the sanctuary. When the snow and the rain ended, we still had over two hundred fifty miniature pigs, along with five farmed pigs, and the right count of dogs and cats. With stability returned and spring now well entrenched, we were even able to selectively take in a few truly desperate little pigs.

One of these emergency calls came from a very irate woman who told me she had a pig on her property whom she wanted shot. She owned a small horse farm in a rather exclusive area, where this abandoned pig had suddenly appeared and would not leave. All attempts to chase him off had proved fruitless, and now the neighbors were complaining.

I calmed the woman down long enough to receive directions to her farm, along with her promise not to kill the pig before I saw him.

I left immediately after feeding the next morning out of fear she would not keep her word. On my arrival, she showed me a small white pig at the far end of her large horse pasture. She insisted he would not come near me, or anyone else.

After walking down the pasture, I squatted before the pig. It was obvious from his appearance and odor that he was a mature boar, though weighing no more than eighty pounds. I began "oofing" to him the way pigs do when meeting one another. His little ears perked up. He trotted over to me and started "oofing" back at a great rate. Though talkative, this little fellow wouldn't come closer than about four feet. All attempts to touch him sent him moonwalking backward out of reach.

After turning up my charm to the maximum on the farm owner, I convinced her that with a little effort, we could capture this pig. I also assured her that I would gladly provide him with a home for the rest of his life.

By now, the woman was intrigued enough by this cagey little guy to help me trap him. After several futile attempts to crowd him into one of her horse stalls, I knew more resources were needed.

Back at the sanctuary, I conscripted Laura. Together, we transported some hog panels to the farm, then constructed a small pen inside a fenced horse run where the pig liked to hang out. I gave the owner a large bucket of mixed feed that all pigs love, then told her: "Put the feed inside the pen and leave open the gate to the horse run. When you think you have him, close him in."

By now, the woman was enjoying the challenges presented by this wily little pig and had become a committed team member. Four days later, we received her triumphant phone call: "He's penned up in the horse run!"

I dropped what I was doing, loaded a cage into our truck, then raced the forty miles to her farm. The woman, along with a friend, was snapping photos of her captured pig, marveling in the adventure. When she expressed concern about how we were going to get the little fugitive from the horse run into the cage, I told her: "This pig knows by now that he will be going home with me. He'll walk right into the cage when I tell him to."

I don't know what prompted me to make such a boast. I felt foolish as I set the cage on the ground, then opened the door. The pig calmly approached the cage. Without hesitation, he walked right in. I closed the door, then grinned at the women as if I wasn't surprised. In truth, I was flabbergasted.

Back at the sanctuary, I placed the little guy, now called Rascal, into the boar pen, then called Tom to perform the necessary neutering. When that was completed, I put Rascal into the downstairs pen with our old guy Elvis to enjoy a few days of recovery before transferring him into the Little Pig Herd.

The next morning, I found our shaggy all-white boy curled up asleep with Elvis in the fresh hay, glad for some pig companionship.

At the same time as we took in Rascal, we fielded a call concerning a pig called Hampton. This one was from the manager of a petting zoo in Hampton, Virginia. It involved another death threat directed at the pig if we didn't "do something." Hampton, I was told, had "gone wild" and had tusked several workers. He was due to be euthanized.

The thought of a small, lonely pig facing death after a life of being poked and prodded by uncaring strangers was too much for me. I told the manager we would take Hampton if he would arrange delivery.

Hampton arrived in a big truck, accompanied by two surly city employees. His cage of heavy plywood, reinforced with two-by-fours and with only a couple of small air holes, would have been well suited for a kicking, charging bull. The two men unloaded the crate, dropped it with no concern for its occupant, then left without a word.

I wrestled the crate onto our truck, then drove it to the pen that I had constructed for our new arrival.

Inside the crate, I found a very scared, very timid, dehydrated, little black and white pig. He had been cooped up for many hours . . . possibly since the day before. I brought him some water, which he drank, and some food, which he refused. My every attempt to touch him made him cower.

Just like Jack Duggan, Hampton had very long tusks, one of which had grown back into his face, where it was deeply embedded. It had been there long enough for skin to have grown around it. This must have been extremely painful as well as made it almost impossible for him to open his jaws to eat.

Once again, I called Tom, signaling to him that this was another tusk emergency. This time, Tom believed me. While I awaited his arrival, I tended to Elvis and Rascal in the downstairs pen. A heavy crash caused me to turn. Big Rosie had pushed her way into the basement, hoping for more food. As I attempted to get her to leave—once again, the equivalent of bantam versus heavyweight—I caught sight of Rascal's white butt disappearing into the darkness. I grabbed a flashlight, then tore into the backyard after him. He had not run far because only about fifty feet away, he had found what he was looking for—an intact female.

Patti Murphy, our seventeen-year-old grande dame, in the midst of one of her now-infrequent heat cycles, had decided to have a drink before retiring into her private quarters for the night. I was just in time to catch Rascal still mounted on her, with the vigor of the stud he still believed himself to be. Poor old Patti appeared to have no idea what was happening to her. She stood with a quizzical look on her face, wondering what this strange white pig was doing attached to her butt—being deflowered in her twilight years!

As I was breaking up this piglet-making session, Tom arrived. He helped me to herd a bewildered Patti to safety, while I could hear Rascal "woofing" his way along the fence, hoping to discover another drowsy, willing female.

But back to Hampton, and the job that still awaited.

I roused the sleepy fellow out of his hay and attempted to squat him on his butt between my legs. My knees weren't what they used to be, so I ended up bear-hugging a squalling, fighting little pig.

Tom sawed off one tusk, while I rolled in the mud, clinging desperately to Hampton. There was no way I was letting go of this pig. Recapturing him in the dark wasn't the way I wished to spend the night. I already had Rascal out there on the loose. While applying wrestling moves that would have made Hulk Hogan jealous, I lost my hat and my glasses, but eventually, Tom had two lower tusks and Hampton didn't have any. The one that had been embedded in his face was a full six inches long.

Tom vaccinated and dewormed Hampton, then continued on his rounds, while I went in futile search of the roaming Rascal.

Hampton spent his first night snuggled in a dogloo full of fresh hay; the next morning, he gobbled a good breakfast . . . possibly the first one in many months that hadn't caused him agonizing pain.

Laura, drinking her early-morning coffee, informed me that she had spotted Rascal in the Little Pig Pasture. I ran out into the dark, with no coat and unlaced boots, to capture him before any intact females wandered out of their barns. Sure enough, in the drizzly morning's gray light, I saw this determined, shaggy white rogue vainly trying to gain access to barns in the Little Pig Pasture. Sleepy and very grouchy defenders were blocking his entrance, unwilling to come out for a tryst.

Slipping and sliding in the mud, I managed to pen Rascal, once again using hog panels, and then wrestled him into a cage. Worn out from his adventure, he slept soundly for the rest of the day.

In the aftermath, Patti was given two Lutalyse shots to eliminate pregnancy just in case Rascal had been quicker and sexier than we thought, with just enough testosterone to complete what he had intended.

As for Hampton, he turned out to be a meek, easily pleased pig who quickly adjusted to yard life, and who would come running whenever a human entered the yard, hoping for belly rubs.

8

You May Be a Sanctuary Director If . . .

During our years at Mini-Pigs, I was frequently asked, "What is a sanctuary director?" and the more probing question, "What does a sanctuary director do?" On the basis of my considerable experience, I will attempt to define the job, both for those who are merely curious and for those who might be tempted to discover the answer the hard way—by actually becoming a sanctuary director.

1. If your personal wardrobe consists almost exclusively of blue jeans, overalls, boots, summer and winter work hats, and a collection of sweat rags . . . you may be a sanctuary director.

2. If virtually all of those clothes have unidentifiable stains, ground-in mud and grass that defy washing, a host of tears, snags, and rips from teeth, hooves, and tusks . . . you may be a sanctuary director.

3. If you have ever had to apply makeup or drastically alter your work or social wardrobe to cover a pig bite, a tusk slash, or a case of red bug and chigger bites . . . you may be a sanctuary director.

4. If the calluses on your hand repel people but make pigs sigh with delight when you rub them . . . you may be a sanctuary director.

5. If your refrigerator is stocked with Ivomec, Dectomax, Pleuroguard 4, Pedialyte (and you have no young children), along with an assortment

of antibiotics, but almost no edible food . . . you may be a sanctuary director.

6. If the medicine chest in your bathroom has one razor and one toothbrush, but shelves full of Blue Kote, Catron IV spray, hemostats, suture kits, bag balm, jars of hoof-grooming treatment, grooming brushes with stiff, black bristles, and jars of stuff marked FOR VETERINARY USE ONLY . . . you may be a sanctuary director.

7. If you open your kitchen cabinets and find fifteen cans of pumpkin, twelve boxes of bran, a dozen bottles of honey, several bottles of dextrose, two bottles of Phenylbutazone, a large bottle of SMZ, but almost no food palatable to humans . . . you may be a sanctuary director.

8. If you can instantly, without using a calculator, estimate a pig's weight within ten pounds, convert that weight to kilograms, then figure out a drug dose based on that pig's weight, but you can't seem to balance your checkbook every month . . . you may be a sanctuary director.

9. If you have ever taken any veterinary medicine because it worked so well on one of your pigs . . . you may be a sanctuary director.

10. If the reading material in your bathroom and on your coffee table consists primarily of *Animal Agenda* magazines from the past five years; books such as *The Veterinary Drug Handbook*, *The Veterinary Merck Handbook*, *The Diseases of Swine*, *Veterinarian Management of the Miniature Pig*; and a wide variety of computer-generated titles such as *Castration Techniques for Cryptorchid Miniature Pigs*, *Diagnosing Swine Endoparasite Populations by Direct Fecal Examination*, and *Home Remedies for Constipated and Impacted Miniature Pigs* . . . you may be a sanctuary director.

11. If your home computer has had all the games and other non-essential programs deleted to make more storage space for pig articles, saved emails, veterinary websites, zoning documents, pig-related court decisions, and pig newsletters . . . you may be a sanctuary director.

12. If you can remember the name, background, arrival date, and in-depth medical information of each of your pigs, including when each intact female comes into heat each month, but you can't remember your

wife's birthday, your anniversary date, or the names of your children . . . you may be a sanctuary director.

13. If you plan your social life around feeding times, farrowing dates of expectant sows, vet visits, feed deliveries, and sales on fencing at the local farmers' co-operative . . . you may be a sanctuary director.

14. If you know the first names of all the guys at the local farmers' co-op and their kids' names, along with other vital information, but can never remember how many grandchildren you have . . . you may be a sanctuary director.

15. If you see more of your veterinarian than of close relatives (including your spouse) . . . you may be a sanctuary director.

16. If you don't hesitate to call your vet at 3 a.m. because a pig is sick but refuse to call your doctor or dial 911 when you experience severe chest pains, or if you refuse to visit emergency when you've been bitten or tusked by a scared pig and are bleeding profusely and obviously need stitches . . . you may be a sanctuary director.

17. If you have ever put on more than one perfume, cologne, or aftershave to hide the unique, musky, and highly obnoxious smell from the preputial diverticulum of a boar you helped neuter earlier that day . . . you may be a sanctuary director.

18. If you have ever used your kitchen table or counter to do surgery, or otherwise to perform any aspect of veterinary care on a pig . . . you may be a sanctuary director.

19. If you have ever had a litter of piglets be born and raised in your house . . . you may be a sanctuary director.

20. If you have ever gone out with family or friends and brought up the subject of how to give an enema to a constipated pig or discussed your latest surgery to remove a uterine tumor and thought nothing of it . . . you may be a sanctuary director.

21. If you have brought pictures of any of the above subjects with you to dinner to share with your friends . . . you may be a sanctuary director.

22. If your last few vacations consisted of attending a pig conference, or visiting or working at a pig sanctuary . . . you may be a sanctuary director.

23. If you plan a four-wheel drive vehicle purchase based on how many cages and carriers the vehicle can hold, how pig-accessible it is, and how big a trailer it will pull . . . you may be a sanctuary director.

24. If you know the ingredient list, protein level, fat content, and the amount of selenium and other vitamins and trace elements of at least four major brands of pig feed but have no idea what was in the food you ate last night for dinner . . . you may be a sanctuary director.

25. If you know the pH of the urine output of each of your pigs but have no idea what your own blood pressure is . . . you may be a sanctuary director.

26. If you have ever picked up pig poop, broken it apart, smelled it, and looked for parasites without giving it a second thought . . . you may be a sanctuary director.

27. If you have done this in the presence of visitors, family members, or close friends . . . you most likely are a sanctuary director.

28. If you can look at a pile of pig poop and tell which pig it came from . . . you may be a sanctuary director.

29. If you've ever put on hold significant home improvements to fund a medical procedure for a pig, or postponed home maintenance to build a structure to house pigs . . . you may be a sanctuary director.

30. If you plan modifications to your home based on how pig-friendly they will be . . . you may be a sanctuary director.

31. If you have ramp access to your home when nobody in your family is handicapped . . . you may be a sanctuary director.

32. If you have ever canceled or backed out of an important social engagement to stay home with a sick pig . . . you may be a sanctuary director.

33. If you have ever spent a night, or nights, in a barn, sleeping with a sick or expectant pig . . . you may be a sanctuary director.

34. If you feel somehow honored that the pigs accepted you for the night, allowing you to sleep with them . . . the chances are even better that you are a sanctuary director.

35. If you can tell the time of day within fifteen minutes based on which tree or bush each pig is sleeping under and which pigs are in the mud hole . . . you may be a sanctuary director.

36. If your computer files and photo albums are overflowing with pictures of rescued pigs, surgical procedures on pigs, and pictures of piglets being born, but you can't seem to find photos of your children or grandchildren anywhere in your house . . . you may be a sanctuary director.

37. If your yard looks like a World War I battlefield in France, complete with shell holes, debris, trenches, and little tiny landmines, and you don't care because the pigs are happy . . . you may be a sanctuary director.

38. If one of your pigs bites the child of a visitor, and you are more concerned about the pig getting a disease from the child than about the welfare of the child . . . you may be a sanctuary director.

39. If you can't understand why parents aren't more concerned about the pig than the child . . . you may be a sanctuary director.

40. If you've ever dragged yourself out of bed, while sick with the flu, during a raging winter storm to give breakfast to pigs who are already twenty pounds overweight . . . you may be a sanctuary director.

Running Mini-Pigs is not an addiction, and it's not a calling. We undertake the tasks, and the ridicule that sometimes goes with owning a pig sanctuary, simply because it is the right thing to do. We revel in the joy of rescuing a pig who has been abused or sentenced to death. We are moved to tears when a scared pig tentatively approaches us for the first time and rolls over for their first belly rub. We grow giddy when a fat-blind, grossly overweight pig begins to see, freed from their stumbling world of darkness and perpetual fear. We worry when a pig is sick or injured, and we struggle to make the right decision when it comes time to consider euthanasia. We grow unreasonably angry when we have to turn away a suffering pig, without being sure whom we are angry at: The "owner" who no longer wants the pig? The rich philanthropic organizations that refuse to fund pig sanctuaries? Our own inability to find room for "just one more?" We become frustrated that we cannot do more, while knowing we are already doing all we can with the resources we have.

Mostly, we just do the work that needs to be done, while enjoying the company of the pigs we have grown to love so much.

9

But Are Our Pigs Happy?

At the height of Mini-Pigs' operation, we had two hundred fifty to three hundred pigs on our seventeen acres, which was substantially more than our normal pig population. Though we had started with potbellies, as the years went by, we welcomed more farmed pigs, having found them a joy to work with. We now had five large pastures, each of which could accommodate thirty-two to one hundred pigs depending on size. Though we allowed our farmed pigs to roam the sanctuary unrestrained, we assigned each new miniature pig to a pasture based on where we had space, like randomly assigning a kid to a classroom.

I knew we were providing all our pigs with a healthy, nutritious diet, proper medical care, and a safe environment, but the question that kept nagging me was this: *Are our pigs happy?*

I loved our pigs. I got such tremendous pleasure from interacting with them that I didn't miss people at all. *Oh God, I was so happy to spend an entire day in a pasture with them.* I got to know all of them—everyone. They are such fascinating creatures, and even more so in groups. I spent an inordinate amount of time observing them, and the more I watched these marvelous creatures, the more that question nagged: *Are our pigs happy?*

After more than a decade of observation, I couldn't in good conscience say that they were. I could not feel confident that Mini-Pigs filled their emotional, physical, and intellectual needs. These very inquisitive creatures didn't have enough to satisfy their insatiable curiosity. After we roused them from their morning slumbers, they would feed, poop, drink water, and then—far too often—they'd lie down under a tree or in the barn and sleep till dinner time. We tried our best to keep grass in the pastures, but during the winter it turned into mud. In the fall, they had a limited opportunity to forage for acorns and hickory nuts, but mostly they were dependent on us.

All pigs have certain traits in common, whether it's a thirty-pound potbelly or a 1,200-pound farmed pig. They are prey animals. They're nomadic grazers and foragers, and they're omnivores. They're probably also one of the world's most hedonistic animals. Pigs live in the here and now. Nobody can enjoy life like a pig can. They suck the joy out of everything with a wanton glee.

If a pig is in pain or otherwise suffering, she doesn't think that the pain will ever end. If I'm putting a pig in a trailer to take him to a veterinary clinic, where he will spend four frightening weeks as strange people do painful things to him, the pig doesn't understand that in a month, he will feel a lot better. All that pig knows is that he's hurting right now, and he doesn't want to hurt anymore.

Anyone who deals one on one with pigs knows they are highly intelligent—ranked fifth to third on the scale of intelligence, along with nonhuman primates, dolphins, and elephants—depending on whom you consult. What I don't believe pigs receive enough credit for is their surprising ability to reason. I've already told you how Patti learned to unlock our chain-fence gate by figuring out what I was doing with my hands, then duplicating my actions with her shoulder and snout. I saw that same ability in our farmed pigs.

When driving the four-wheeler through the barn gate, I would distract lurking pigs by tossing a few apples. This was way easier than wasting half an hour chasing down escapees as they romped outside of the pasture. After parking, I would then take the rest of the apples out of a bushel in the back of the four-wheeler and distribute them among the eagerly waiting pigs. That done, I would use the latch underneath the four-wheeler to release the back dump bed to the downward position to hose it out.

After my farmed pigs had watched me do this a couple of times, they figured out not only the sequence but also how to turn it around to their advantage. While I was still parking my truck, I was startled to see Rosie and Big Boy—two huge one-thousand–pound Hampshires—walk up to the back of the four-wheeler, find the latch, then release it with their snouts while pushing up on the bed with their shoulders. This neatly dumped the entire bushel of apples in front of them, which they happily devoured, having eliminated me as the middle man.

More than any other animal I have ever known, pigs also have a desperate need to be part of a social group. Not only do they enjoy the companionship of other pigs, but they also have strong feelings about which pigs they like and which pigs they don't like, apparently based on their judgment of character.

I'm fond of a very old Irish poem, author unknown, that slyly makes clear how choosey pigs can be about the company they keep. I wish I could recite it in a lilting Connemara brogue, but this will have to do:

The Irish Pig

'Twas an evening last November,
As I very well remember,
I was strolling down the street in drunken pride,
But my knees were all a-flutter
And I landed in the gutter,
And a pig came up and lay down by my side.

As we wallowed by the curbing,
Not a soul were we disturbing,
When a lady from the church was heard to say,
"Ye can tell a man that boozes
By the company he chooses."
And at that, the pig got up and walked away.

I had routinely been putting together pigs who didn't choose to associate with each other in a confined space, without an opportunity to walk away. Just like family members who don't particularly like each other, they got on each other's nerves. While a couple might pair up, group friendships were the exception rather than the rule. Since pig societies are matriarchal, that bonding usually began with a female alpha pig who would attract a couple of other pigs to hang out with her, like a queen with ladies-in-waiting.

In most of our pasture herds, the undercurrent of unease was just one level below turmoil. A lot of disagreements broke out, a lot of fights, followed by a lot of moving between barns.

Most pig fighting was done one on one, with a great deal of vocalizing and posturing. When this escalated into a physical challenge, one pig might hit the other, shoulder to shoulder, or put their snout under the other's chest and try to throw the other.

Though female fighting was not as frequent as male fighting, the females would fight harder and more viciously, with more biting of ears and tails in a way that might tear flesh or draw blood. This was always about dominance, with an alpha female putting down a young pig trying to fight her way up the social ladder. If I noticed one pig picking on another, I could predict that within a couple of days, those two would be fighting.

For the females, this was usually a one-time event. They'd fight for a few minutes, rarely more than five to ten. One pig would give ground or break and run, and the issue would be settled once and for all. That night, the two former combatants could often be found sleeping together. Pigs don't hold grudges.

It took me longer to understand male fighting since we neutered boars on or before their arrival at the sanctuary, limiting our experience with male conflict. That is, until we rescued Sergeant Major Tinker, a fifteen-year-old cryptorchid (a male with one retained testicle). When our vet decided Tinker was too old to be a good candidate for surgical removal of that hidden testicle, we were left with a dilemma. We couldn't put an intact male pig into the general population because not all our females were spayed. On the other hand, I didn't want to isolate Tinker in a pen by himself for the rest of his life.

The solution appeared soon afterwards when we acquired two young intact males, Trooper and Oogie. We delayed neutering them so they could become companions for Tinker. I designed a special pen with three layers of fencing, a barnyard version of Fort Knox in terms of security. I did not want our perfect record of no accidental pregnancy to fall thanks to the testosterone of a couple of young males.

Each day, these three boars would get up in the morning, stretch, eat, poop, drink some water, then start fighting. It was like watching three sumo wrestlers. They would plow the ground with their snouts, then do this little Michael Jackson moonwalk with their front feet. They'd foam at the mouth and toss each other around for two or three hours until they were exhausted. Then they'd take a break, maybe drink more water, and lie down together for a nap. After a while, they would get up and start the process all over again.

This behavior went on all day long, every day. Unlike with females, who fought to establish herd dominance, it seemed to be more akin to a sport for these three males, with no desire to lead the herd. Though they possessed much longer tusks than did the females, nature had also provided them with a thick coat of armor-like skin that went from just below their ears down to their ribcages and front legs so that each resembled a little rhinoceros. Since this skin was probably three-quarters of an inch thick, their tusks didn't usually cause serious injuries when they tussled.

Of course, our sanctuary was not a natural environment, since intact males are genetically programmed for two things: eating and mating, not necessarily in that order. Everything they do is geared towards those two goals. In a natural environment, the boars must fight to mate because the dominant females only allow dominant males to breed with them. Once impregnated, the females then chase off the males so they can farrow and collectively raise their young. They don't like intact males around because they often kill the piglets.

A sanctuary is a reflection of its owners—a lot of one's personality and beliefs go into it, so no two are alike. For me, saving a pig from death, putting him in a pen with a bunch of other pigs, then feeding and caring for him for the rest of his life were not enough. I wanted to give our pigs as independent and as natural a life as possible.

I wanted our pigs to be happy.

To plan an environment and a lifestyle for an animal so similar to—and yet so different from—us meant accommodating both similarities and differences. I had to stop looking at a pig through human eyes and start looking at the world through pig eyes. That was a difficult process, but it yielded positive results.

Pigs have a lot of abilities we don't have, and we have abilities that pigs don't have. I can't smell an acorn or an apple a hundred yards away like a pig can, but I can grow and harvest acorns and apples or buy them at a store. Our two species also possess different desires. I would not consider wallowing in a mud hole on a hot summer's day to be a pleasurable experience, yet for a pig that mud hole is a sensuous spa. I would not by choice sleep in a barn full of itchy hay, but for a pig that's a heavenly spot in which to snooze.

Over the years, every facet of our sanctuary caretaking became shaped by my ability to step out of my humanness and to put myself inside the skin of a pig. My observations were aided by our farmed pigs, who were free to roam, allowing me to see how a pig entertained himself when left to his own devices, compared with how a pig behaved when fenced in, with nothing to do except be with other pigs.

The more I learned about the emotional, physical, and intellectual needs of our pigs, the more I respected them, the more I wanted them to be happy, and the more I realized that our seventeen acres in northern Virginia fell woefully short of fulfilling those needs. I wanted to create a natural preserve, where the pigs would have as much independence as possible while we continued to take care of their veterinary needs. Some rescue groups had achieved this for elephants and primates, but when I spoke about adapting that concept for pigs, the experts said this was not doable. Pigs could not live in nature because their natural instincts had been genetically engineered so we could domesticate them as farmed animals. In short, they were no longer "real pigs." If we transitioned from the traditional sanctuary model, where pigs are contained and constantly supervised and monitored, we would be creating a situation that allowed for survival of only the fittest.

Since these were some of the same experts who had warned that miniature pigs and farmed pigs could not possibly get along together, I did not believe them when they said that giving our pigs the gift of freedom would doom the weakest among them. As a counter, I carried in my mind the image of Big Earl and little Scooter ambling side by side to their mud hole, preparing for the pleasures of the day.

When I discussed with Laura the idea of evolving from a traditional sanctuary to a preserve, I came to realize that this was more my dream than Laura's. While I bore the brunt of caring for the pigs, I was also the one who had the pleasure of their company. Laura's and my relationship had suffered through the demands of holding down two jobs while also running Mini-Pigs. We were always sleep-deprived. We never had time to carve out for activities as a couple. We had few mutual friends. Our basic forms of communication sometimes narrowed down to phone messages and post-it notes. Our only vacation together had been our ten-day honeymoon, which we had cut short by five days to get back to the pigs. Though it seemed that Laura might be reaching her saturation point, she agreed to support me the way she always faced challenges—100 percent.

That opened up the question: How could we accomplish the transition?

From the founding of Mini-Pigs, Laura and I had figured that the best way for an operation like ours to succeed was to liaise with other mom-and-pop sanctuaries. Through the years, we had made a massive effort to reach out to the few other small sanctuaries and rescue groups that were popping up around the country. This included a group in Florida called Pigs as Pets, run by Lana Hollenbeck, a longtime friend. Though Lana didn't have a sanctuary of her own, her organization helped fund rescues and bring sanctuaries together. This networking started with a newsletter and the establishment of a small board of directors. Though it was a modest effort, the exchange of ideas provided both practical and psychological support to our diverse pig-rescue community.

When we decided to evolve from a sanctuary to a natural preserve, all of our liaison work paid off in a useful way. For the past couple of years, we had been downsizing our herd from more than two hundred pigs to around one hundred. This was accomplished partly through attrition, by not replacing our older pigs who died, and partly through our decision to move from rescuing miniature breeds to farmed pigs. Several potbelly sanctuaries had sprung up in Virginia that we had mentored through advice and a mutual exchange of physical help on weekends. This made us feel more comfortable about not rescuing every potbelly we were offered. It also meant that when we wished to downsize in preparation for moving, these sanctuaries were willing to take in some of our potbellies.

I should explain that our "downsizing" referred to numbers, not to bulk. I would say to another sanctuary director, "Look, I'll take your farmed pig if you take four or five of my older miniatures." In terms of dollars spent on upkeep and space requirements, a 1,200-pound farmed pig was equivalent to about ten 150-pound potbellies.

While Laura and I were negotiating these trades, we were looking for a larger piece of land that would eventually become The Pig Preserve. This process began with a laundry list of features we didn't like about our current location, matched with a laundry list of features we wished our new location would possess in a perfect world.

One of our first desires was for four distinct seasons, not brutal winters and god-awful summers, because that was too tough on us and on the animals. We did not want to be on the east coast because of the ever-present need to evacuate for hurricanes. We didn't want to be too far south or west because we did not want to be in tornado alley. We didn't want any place prone to earthquakes or raging forest fires. We didn't want a whole host of poisonous plants on the property. We didn't want to be in any state with overbearing regulations or complex zoning ordinances that might haunt us in the future.

We did want multiple sources of natural water, with spring-fed ponds for our pigs. We also wanted the right mix of pastures and woods. We wanted hills with good drainage but not so steep that our pigs would have difficulty negotiating them. We wanted good neighbors. We wanted a place at the end of a road with no through traffic. We also wanted the biggest bang for our buck.

Curiously, given our fairly stringent set of criteria, the house where we would live came in at around number eighteen on our wish list.

Because Virginia was fast becoming too expensive, we started looking out of state. We settled on the Kentucky–Tennessee area, then narrowed that down to Tennessee. It had four seasons, with a more moderate climate in the mountains. It had no state income tax, moderate property taxes, and not too many nonsensical regulations about building codes and vehicle inspections. Quite a lot of rural land was for sale, and it was cheap, fulfilling the "bang for our buck" requirement.

After extensive juggling, we found temporary placements for all of the animals we intended to keep, some being left with the woman who had purchased our Virginia property. Then, in 2005, Laura and I took a huge leap of faith. We both quit our jobs and moved to Tennessee, officially homeless.

While I felt sad about retiring from the work I loved with the Fairfax fire department, I knew it was time. Because of my previous military career, I had been one of the oldest people to be hired as firemen. My twenty professional years with Fairfax, added to my ten previous years as a volunteer, had taken a toll on my body. Though I was still in relatively good physical shape, I was having problems with my knees and my back. I was dealing with the first stages of arthritis in a number of joints, and I had lost count of the number of sprains, strains, and bone fractures I had suffered over the years. Firefighting is the ultimate team sport. Every crew member has to be sure that every other crew member is capable of rescuing them from a life-threatening situation. I started to lose confidence in my ability to do that. As I aged, I knew I would feel like more a liability than an asset to my fellow firefighters. My department allowed retirement with no financial penalty at age fifty-five, after at least twenty years of service. I was fifty-seven. I had that option and I took it, with more sorrow than relief.

Laura enjoyed her work as a hospital administrator and was not ready to retire. Since both her qualifications and her job were universal, she knew she could find work at another hospital without much trouble. She also happened to be really good at what she did, which involved insurance, coding, and paying bills.

For about three months, Laura and I lived in a motel with two dogs and two cats while we searched frantically for a place for our new sanctuary/ preserve. We probably looked at a hundred different properties, with checklist in hand. Some had beautiful homes, even guest houses with shops, manicured lawns, and all the amenities. If the land wasn't suitable for our pigs, we passed it by.

Finally, we found a property in Jamestown that met 99 percent of our criteria: roughly one hundred acres of rolling hills, including forty acres of lush pastures, four spring-fed ponds, and sixty acres of woods full of maples and nut trees. Located on a dead-end rural road, it had been purchased and cleared about sixty years ago by a local man who had used it for cattle farming. His widow, who sold the place to us, had let it go to seed over the fifty years since his death. The farmhouse—small, single-story, with a full basement—was in need of some tender loving care, but the land was perfect for the pigs.

Before looking for work in Tennessee, Laura took a year off so that the two of us could spend twelve to fourteen hours a day, seven days a week, preparing our new property for our pigs. This was a massive upfront undertaking. With money left over from the sale of our Virginia property, we purchased a second large tractor with all the necessary attachments: post-hole diggers, mowers, rakes, discs, forks, and buckets. We also purchased a diesel mower and two well-used large diesel four-wheel drive pickup trucks, along with a couple of trailers—all told, a sizable investment for two jobless people starting a major project from scratch rather late in life.

Our first task was to clean up the old barn that came with the house—a forty-by-sixty–foot cinderblock and wood building with two stories, including a hayloft. Far more challenging was replacing miles and miles of old, rusty barbed wire with adequate pig fencing. The existing fences were probably sixty years old, and if you've ever gathered up old barbed wire, you'd find it's like wrestling with porcupines. We must have gone through a box of Band-Aids every day. Then, we had to haul truckloads of the stuff away. Next, we ran fencing suitable for pigs, which meant handling sixteen-foot sheets of cattle panel, fifty-two inches high, mostly following the boundaries of the old pastures. To accomplish all this, we'd wake up at 6 a.m., grab a quick cup of coffee, load the tractor and hay wagon with fencing materials, then work until dark.

We also had to cart off truckload after truckload of garbage—old tires and rusted machinery—that people had for decades been tossing into our woods. Finally, we had to arrange a rescue for a horse abandoned on our property.

In Virginia, we had used enclosed barns of varying sizes and styles. Nevertheless, each winter, we had had pigs suffer from pneumonia, bronchitis, and a host of other respiratory ailments. Perplexed, I sought the advice of some farmers who raised pigs in a natural setting. While I didn't approve of *why* they were raising pigs, I figured they would know, if anyone did, how to maintain a healthy herd.

Without fail, these farmers informed me that long, cold winters would not be a problem for our pigs so long as they could be kept dry and out of the wind, and provided with adequate hay or straw as bedding. In fact, they assured me that pigs living in "open barns" rarely suffered from any diseases, including the numerous respiratory ones we'd dealt with over the years.

Taking this guidance to heart, Laura and I built a couple of permanent three-sided shelters, to which we added three portable twelve-by-twenty-four–foot metal sheds on skids, as our first pig shelters.

The old farmhouse was a story unto itself—a derelict that lived up (or down) to its placement as number eighteen on our priority list.

Its flat-roofed cinderblock basement had once been where the family had lived. Eventually, they added the upstairs, which still lacked air-conditioning and heating. We decided we could live with it. And we did. Continuously. Because every time we saved enough money to fix a problem, a pig emergency claimed the funds. Fixing our old metal roof was high on our to-do list. As I write this, it still leaks, while the new roofs on our pig barns assure that our pigs will never get wet.

Life is a matter of priorities.

Over three or four months, I made thirty-six trips transporting pigs and equipment the five hundred miles from Culpeper, Virginia, to our Jamestown property. If the weather was bad, I drove my flatbed trailer to haul fencing and water troughs and everything else needed for our sanctuary. If the weather was good, I hitched up the stock trailer and transported a load of pigs.

Surprisingly, the pigs were no trouble to transport, partly because it was summertime. After pulling into Virginia in early afternoon, I would grab a few hours of sleep. Then, late in the evening, after the pigs had eaten and drunk and pooped, I would load about twenty of them using the temporary chutes we had built. I'd leave at midnight, with the pigs asleep in the straw in the back of the trailer. I'd drive through the night—about ten hours—to arrive in Tennessee at daybreak. It was like a red-eye special, at least for me. Most of the time, the pigs didn't even wake up between their old home and their new one.

What happened next amazed me.

I had thought the pigs would need some breaking-in period, during which they would feel overwhelmed by all the free space and perhaps a little afraid to venture out into their new world.

There was no learning curve at all.

Without fail, whether it was potbellies or farmed pigs, when I opened the trailer door, about twenty eager pigs would run out. Within three minutes, we wouldn't see a single one. They'd all be down in the woods or splashing in one of our four ponds.

It was as if they had been born to it, and genetically, they had.

I could already imagine the future—two hundred pigs, their prior lives of trauma, abandonment, neglect, and abuse now forgotten, running free, foraging, and exploring, living on their own terms, just being pigs. It seemed like our gamble had paid off—or was I being too romantic? Was I still seeing the dream and not living the reality?

THE PRESERVE

10

The Best of Friends

EVEN THOUGH LAURA AND I had extensively researched the preserve concept, we still had trepidations. This was a big gamble for us. When we left Virginia, we had shut down Mini-Pigs, which meant starting anew. In 2006, we created The Pig Preserve—a tax-exempt, nonprofit sanctuary, which would be the largest pig-only sanctuary in the US. Though our board of directors had dwindled down to Laura, me, and one other person, that was all the state of Tennessee required.

As predicted, Laura easily found a job in a local Tennessee hospital, doing much the same work as she had in northern Virginia, albeit for a significantly lower salary.

I volunteered with the East Jamestown Fire Station in Fentress County, where I was made a station captain based on my training and experience. This was something I would continue to do for the next ten years, until my heart started giving me problems and my doctor said I should quit. While our small, rural firehouse didn't provide the frantic pace I had previously learned to live with and even enjoy, I participated in responding to the occasional fire, auto accident, or other emergency.

Even today, I miss the exhilaration and adrenaline rush of firefighting. Firemen are a lot like little boys. We thrive on the challenge and the thrill. Though caring for pigs never provided me with that daily injection of excitement, it gave me a covering blanket of pleasure, serenity, and deep satisfaction. I never missed the paramedic side of emergency work because I had grown too callous. My proficiency level was still right up where it should be, but the emotional gap had grown too wide, which is a true sign of burnout.

At Mini-Pigs, I had observed the rudimentary formation of friendships among our pigs. Now, I was curious to see what more than one hundred intelligent pigs of varying breeds, ages, and sizes would do when given the opportunity to freely associate without the constraint of fences. Would the herd turn wild and aggressive at the expense of elderly and disabled pigs like Big Rosie and Babe, as some experts had warned? Would Hazel and Ella be pushed back into the private worlds of trauma from which they had been rescued?

For about four months, the pigs roamed the perimeter of the property, gauging the size, scope, and diversity of their new world. When they collected in groups, these seemed random rather than chosen. Eventually, I began to notice a series of fights, or discussions, if you will: A small group had formed that another pig wanted to join. After the interloper was driven off several times, they might still return, insisting on inclusion. With each effort, the fighting would lessen until that particular pig was accepted. A group of three would become a group of four, then five, then maybe even ten.

At some point, that group would drift off from the herd to a particular part of the preserve that appealed to its members—maybe by a pond or in a back pasture. After noting this choice, I would put one of our twelve-by-twenty-four–foot barns for them in their chosen spot. In this way, each new social group became a little community with its own housing, its own food, its own access to water, and its own area to play and forage.

After about six months, these groups began popping up all over the sanctuary. By the time we had one hundred fifty pigs, we had about twenty individual communities. We knew this wasn't going to be a static situation because some of our older pigs were dying while we were bringing in new rescues at a slow but steady pace. Also, the social groups themselves had a certain fluidity. A squabble would occur in what had seemed a tight group, and one or two pigs would leave. They might wander around for a month or so until they were accepted into another social group that they now seemed to prefer. Since we had built all of our barns on skids, we could move an empty one to where a new group was gathering.

As the seasons passed, the social habits of our pigs also evolved. Our rescues in spring and summer didn't feel as much pressure to join a social group because of the pleasant weather. Instead, they—especially the younger ones—might sleep out in the woods or down by a pond. Some pigs already in groups might also decide to abandon their barns, so I had to go into the woods searching for them. In the fall, when temperatures cooled, the pigs roaming free would realize they had better find a social group in order to have access to a barn. Some might return to the group they'd left to see if it was still to their liking. Newcomers would be looking for a social group—and a barn—for the first time.

Because of all this shuttling around, within and without our barns, Laura and I learned not to take in too many pigs in the dead of winter. It was just too hard on us and too hard on the pigs. After October or November, we accepted only those facing life-or-death emergencies, with all other needy pigs being placed in foster homes till spring, when we could introduce them to the sanctuary a few at a time.

I had figured that when left to their own devices, the farmed pigs would gravitate to other farmed pigs, potbellies to potbellies, the young to the young, and the old to the old. That's not what happened. The pigs chose their friends according to their personalities. They either liked or tolerated each other without much regard to age, sex, color, size, or breed.

Our miniatures remained the most argumentative of all of our pigs. They loved to fight, both among themselves and with any other pig they happened to encounter. Every minor issue required a great deal of squealing and other histrionics before being settled. Whenever I heard a disturbance in one of our barns, I would bet my money that it had been started by one of our willful, headstrong miniatures. Sometimes, they formed roving gangs on the lookout for a bit of devilment. Though they could be a nuisance, it was hard not to love their spunk and spirit.

Fortunately, our laid-back farmed pigs treated the scrappy miniatures with tolerance. When one picked a fight, the much larger pig would most often back off, then nonchalantly amble away. If a tenacious little pig went too far, the farmed pig might use their massive head to flip the offending miniature head over heels, but I've never seen one of these giants intentionally try to hurt a smaller pig.

In winter, our rambunctious miniatures reversed their tactics. As the temperatures dropped, they loved to bury themselves in the hay under the massive farmed pigs in order to siphon off some body heat. Again, our gentle farmed pigs, like Timbo and Big Boy, not only accepted these pesky heat thieves but were careful not to crush or injure them. Timbo—white, with the long snout and heavy drooping ears typical of his Landrace breed—had spent the first part of his life as a breeder boar for a small farmer. The farmer had noticed that Timbo was so tender-hearted that he grieved terribly when each litter of piglets he had sired was shipped off to be slaughtered. After the farmer got out of the pig business, he retired this sensitive fellow instead of slaughtering him. At the preserve, Timbo easily found more than enough piglets and potbellies to satisfy his compassionate heart. The same was true of Big Boy, a very large Hampshire who sought out potbellies as pals.

Pigs who came to us as singles usually had the most trouble adjusting to a natural environment in which they were free to just be pigs. Typically, they had been ripped from their mothers and siblings at a very young age, without a chance to learn how to be a pig. To them, the other pigs were frightening aliens. If they had learned to bond with humans, they might attempt to fulfill their social needs by creating a surrogate herd out of me and our volunteers. When we were working in a field, these confused pigs would sometimes become pests, taking our tools or otherwise drawing attention to themselves. Turning them away was hard. They were so obviously needy, but giving in to them only made their acclimatization more difficult. They had to learn to fulfill their emotional needs with other pigs.

Some pigs who came to us malnourished would follow us around out of fear they weren't going to be fed. Once they had full bellies on a regular basis, they learned to forage with the rest.

New rescues who'd been raised with animals other than pigs might attempt to create a surrogate herd out of our dogs and cats. While that wasn't as fulfilling as bonding with their own kind, it felt better to them than being alone.

Younger pigs typically had an easier time than did older pigs, because they had less history to undo and they were more mobile and inquisitive. Those who came to us in small groups had the easiest transition—like the three Berkshire brothers, Wynken, Blynken, and Nod. Though they arrived emaciated and covered in parasites from life in the wild, they were already open to having a good time.

Because the farmed pigs had so often spent their whole lives in confined areas, their leg muscles and joints had often atrophied, making them slower to appreciate the possibilities of their big new life—lolling in the cool mud of a spring-fed pond, gorging on fall blackberries, competing for apples, and falling asleep beside a best buddy.

One of our saddest rescues was Miss Sugarfoot, who had been locked in a dark barn for over two years without attention or proper food. Because her hooves had grown to over eight inches in length, she couldn't walk. Instead, she flopped on her side then dragged herself along the ground. As a result, she had no hair and her skin was rubbed raw.

Our vet believed Miss Sugarfoot would never walk again due to the damage to her feet and dewclaws (digits on the outer side of a pig's hooves). After months of rehabilitation, we placed her with a small herd of crippled pigs in a fairly large pasture. This allowed her more room to roam, helping to keep her recently trimmed hooves worn down. She gained weight, grew a new coat of hair, and yes, she did learn to walk, albeit with a bad limp from arthritis in her front feet. She also found pig companionship for the first time in what had been a nasty life.

Despite all my experience of tragedy, I was not prepared for what greeted us that day in 2014 after we had accepted a transport of three mature farmed pigs, ages unknown, from a crisis situation in Florida. When I opened the trailer gate, out staggered three pitiful skeletons. You could see every bone in their bodies. I broke down in tears, astounded they were still alive.

Authorities had found these siblings at a hidden, illegal slaughterhouse, where they had been abandoned along with a host of other animals. Many in this hell hole had already been dead, and a number had had to be euthanized.

After we settled them into their own barn, our vet informed us that she didn't believe they would survive. More out of hope than expectation, I made soft, palatable meals for them, relying heavily on cooked rice, oatmeal, and bananas, mixed with a small amount of ground whole grains. Because their stomachs were so badly shrunken, I had to hand-feed them six to eight times each day, as well as constantly syringe water into them.

Despite having suffered terrible abuse at the hands of humans, Rufus, Ruby, and Rhonda quietly allowed me to feed, water, and care for them. Over weeks, and then months, I doctored wounds, drained abscesses, treated them for a horrific mange infestation, all the while helping them to feel comfortable in the presence of a human who didn't want to hurt them. With serene dignity, they accepted my care. Slowly, I was able to reduce their daily feedings from eight to two, while also introducing them to a more normal diet of mixed grains, molasses, and vegetable oil, supplemented with children's chewable vitamins. In summer, I set fans in their barn to help keep them cool and misted them with water from a backyard sprayer on the hottest days.

Ever so slowly, the appetites of these three siblings returned, their coats began to grow back, and a spark of life returned to their eyes. By the end of August, Rufus, Ruby, and Rhonda were standing, walking, and showing a definite interest in getting out of their barn. This steady improvement was very gratifying to me.

In the early fall, I was comfortable enough with their physical progress to allow the three siblings access to the main herd. That was when their dormant but strong social preferences took hold. Ruby and Rhonda spent a day and a night wandering around before settling in with a mixed group of potbellies and younger farmed pigs.

Despite all Rufus had been through with his two sisters, he immediately left them to move into the main cinderblock barn, where he became fast friends with Rosebud, Ciara, and Rosie—three older farmed girls, nicknamed "The Cougars" for their unusual affection and affinity for younger male farmed pigs.

After Ciara and Rosie passed on, Rufus—who grew to a svelte one thousand pounds—remained tightly bonded with Rosebud. At last count, they had collected around them ten potbellies of both sexes, ranging in size and age, and a couple of two-year-old farmed pigs. Each night, two other potbellies who had been living in another area of the sanctuary tried to sneak into the barn. They'd hang around after feeding time until the group had gone to bed. Then, they'd tiptoe inside, sleep on the periphery, and scamper out at daybreak. Each night, they would sneak in a little sooner and stay a little later. By the time winter set in, they were accepted.

Rufus and Rosebud—now elderly and arthritic, with limited mobility—wanted to sleep more soundly than the young pigs, who would awaken early or even in the middle of the night. The youngsters were noisy. They were playful. Rufus would grumpily chase them outside, then go back and lie down, or maybe he'd help Rosebud up and they'd amble to the other side of the barn. It was a pig version of Sunday morning when mom and dad would chase out the kids so they could sleep in: "Go watch TV or something!"

Both Rufus and Rosebud were on daily medication—hidden in giant peanut butter and jelly sandwiches—to reduce their joint swelling and mitigate their discomfort. Yet both managed to get up at least once a day to totter outside to pee, poop, get a long drink of water, perhaps even walk around for a few minutes. Occasionally, Rufus rooted himself into a big hole in the hay and needed human assistance to get up. He let us know, very vocally, when he was stuck, then waited patiently for us to boost him back onto his feet.

The strong sociability patterns we found among farmed and potbelly pigs also applied to the feral pigs whom The Pig Preserve began to accept. Though ferals are commonly called "wild boars," that is a misnomer. It is also incorrect to use the term "feral" for domesticated pigs of other breeds who have escaped from captivity and have lived in the wild for several generations. Ferals are a third, unique breed of pigs, typically three to five hundred pounds at maturity, with thick, unruly hair and a "Mohawk" or "razorback" down their spines. Their snouts are long and angular, and they have very pronounced tusks. They are much more agile, more muscular, more energetic than either farmed pigs or miniatures.

Ferals in nature prefer to stay as far away from humans as possible and are frequently nocturnal. They live in traditional herds or "sounders," with an alpha female in charge. They are hardy and very adaptable, able to exist by rooting and foraging. They are also nomadic, often traveling many miles each day, then making fresh, family-sized nests in the underbrush each night. Over the past few years, they have received a great deal of adverse and unwarranted publicity thanks to the media's portrayal of them as ferocious, human-hunting beasts. Nothing could be further from the truth. Like all other pigs, they are shy, retiring creatures who would much rather flee a confrontation than fight. However, they will react viciously if scared and cornered, or if they are protecting their families.

Feral pigs are the genetic ancestors of European and Russian pigs, brought to this country by explorers as early as the 1500s. Those who broke loose from captivity migrated into the coastal swamps, especially in the Southern and Gulf Coast states, where they evolved into the animals we see today. Since the weather was temperate, and they had plenty of food and few natural predators, they thrived. Even humans mostly left them alone.

Southern ferals didn't suddenly decide to head north to Tennessee or Michigan or Minnesota, or any of the other states where they are now considered a nuisance. This was a human-made problem. They were trapped in the Gulf Coast area, then transported north—illegally and without blood tests—by people who ran "canned hunting camps." These were confined pastures where clients could shoot pigs and other wild "game" for "sport." Many of these wily and resourceful pigs escaped by tearing down or rooting under perimeter fences. When these camps went bankrupt, the owners released the rest of the ferals into the woods, where they multiplied.

Even though ferals look extremely ferocious, they are still prey animals, relatively defenseless against predators. While humans are their toughest adversaries, they also have reason to fear big cats and coyotes. Based on their reality of having been hunted in the wild, ferals arrive at the preserve with a strong preference for the out-of-doors, even in winter. Often, they will dig a cave, then drag in brush and leaves to make a nest as they did in nature. Once ferals succumb to the indoor comfort of sleeping overnight in a barn, they are still the first pigs up each morning and the last ones to go to bed. When their fear of human touch breaks down, they, too, became suckers for belly rubs and ear scratches. Their inexhaustible energy makes them fun to have around. They love to run, jump, and play. Even in winter, you will see them frisking about in the snow. I have to be careful when playing with them because of their impressive, razor-sharp tusks, which can inflict a serious wound.

Since the state of Virginia didn't have much of a feral problem, we had had only our inbred Ossabaw Island group at Mini-Pigs, along with a few crossbreeds: feral–farmed mixes, feral–potbelly mixes, and feral–farmed–potbelly mixes. In Tennessee, ferals are numerous enough to be considered destructive pests. Though it is illegal to own one here, The Pig Preserve has been given special permission from our state veterinarian to rescue and care for them.

Our preserve's first true feral was a magnificent fellow named Chopper Harley Hog. As typical of his breed, he was a scary-looking pig, well over five hundred pounds, with a razorback, huge tusks, and a fearsome demeanor. When ferals are neutered young, they grow up soft and round and eunuch-like. Chopper had been an intact male much of his adult life, so he retained his boar-like characteristics. When you first saw him, it was difficult not to be intimidated.

I wish I had a reliable backstory on Chopper, because right from the start, he defied all feral expectations by being one of the easiest, most playful pigs we've ever had. His appearance was so terrifying that if he approached a fence, people would instinctively draw back. As part of a tease for visitors, I would put an apple in my mouth, then let Chopper take it ever so gently from me, always being careful not to cause injury.

Chopper moved in with a group of young farmed pigs of probably two or three years old, and that group more or less stayed together until he died. He and I had such an easygoing relationship that I commemorated him with a caricature tattooed on my arm.

We had another effortless time with a feral named Silico. In 2016, a young couple was driving outside of Chattanooga when they saw what they thought was a dead possum. After the creature moved, they pulled over. It was a feral pig of probably no more than a couple of days old, covered with hundreds of ticks and too weak to stand. A local vet removed the ticks, got some fluids and nutrition into the piglet, then warned the couple that it was illegal in Tennessee to own a feral. If authorities heard about it, they would confiscate and destroy the baby.

When the couple called me, I told them we were overwhelmed by piglets, but if they could hang on to him until he was about twenty to twenty-five pounds, then we'd take him.

The couple raised the pig in their apartment by bottle-feeding him until he was twenty pounds. By then, I had created the Baby Pasture, so I told them to bring him right over. Silico, who had never been outside their apartment, looked like an oversized chipmunk, with his light brown hair and spots. When they put him down on the grass, he took one quick look around, then beelined for the nearest mud hole, nosedived in, and began flopping around.

Silico fit right in with a bunch of farmed pig youngsters, then grew to be three hundred pounds with an impressive razorback. He became a super-friendly delight, running to visitors whenever he saw them, eager to get his belly rubs or to score a treat. He also liked to "nurse" on our clothes—a trait found in many pigs taken too early from their mothers. Usually, they prefer to suck on fingers, but one of our feral–farmed pig mixes named Twiggy liked to vigorously suckle knees. This was considered cute until she grew to be over two hundred pounds. With maturity and mild correction on our part, she eventually lost this urge.

Damien and Sunshine, two beautiful adult feral pigs, came to us from a potbelly sanctuary in Florida where they didn't have enough space to roam. What farmed pigs or potbellies experience as a fear of confinement, ferals experience as absolute terror. Transporting these two guys to the preserve cost thousands of dollars in trailer damage. On their arrival, they set up camp with a group of farmed pigs while still retaining their own close bond. Though they grew comfortable around any human they came to know, they remained shy around visitors.

I have seldom seen two pigs as closely bonded as Damien and Sunshine. After they aged into their twenties, they became less competitive at the crowded feeding troughs and began to lose weight. In addition, we began to suspect Sunshine was having problems with her vision. Not only did she constantly stay by Damien's side, but her eyes appeared to be clouding over as if she might be getting cataracts.

To make life easier for this geriatric couple, we moved them to our approximately fifty-by-fifty–feet "quarantine pen," with their own small barn and a private stock tank with a ground-level "pig drinker."

Sunshine and Damien took to their new home instantly. Both regained their lost weight, and for two more years, they lived in comfort and serenity. After that, we suddenly lost Sunshine to what we believe was a heart attack. Sadly, our concern that Damien would grieve himself to death proved true.

Journey and her seven siblings were only hours old when hunters killed their feral mom, then let their dogs ravage through the baby pack. Journey still had her umbilical cord attached to her belly when the hunters decided to keep her, probably intending to use her as bait to train hunting dogs. Within a few days, a rescue group found out about this situation and managed to save her. At the preserve, Journey became friends with Luna Rose, a two-week-old, three-pound farmed piglet tossed into the dead bucket at a factory farm because of her small size. Her rescuers named her for the full moon on the night of her rescue and in memory of our Big Rosie, who had just died.

Because of the special permission The Pig Preserve has to care for ferals, most of these rescues came to us from state and county animal shelters. Some were saved by hunters who took pity on babies they found trying to survive in the wild after their mothers had been shot. Once animal control became involved in a feral pig rescue, their staff generally didn't want the adverse publicity that would come from shooting or gassing a group of cute, innocent piglets. This was when our phone would start ringing.

One group of four ferals was transported to us from an animal shelter in Memphis. We named them Memphis, Martin Luther, Coretta Scott, and King. Though all appeared to be very young, about half a day after their arrival, Memphis surprised us by producing seven babies. Our four-pig rescue had become an eleven-pig rescue!

We dropped everything to construct a piglet-proofed fence around the new barn where the young ferals were living. Since cattle panels, with their openings, could not contain these babies, each weighing less than a pound, we ran a second fence of chicken wire inside the primary one. While this did an excellent job of keeping in the new family, a fifty-two–inch fence would not keep out any coyote who happened to be in this remote area. Our solution was both simple and old-fashioned. Since coyotes and other wild predators generally avoid areas with a strong human scent, every evening during feeding, or any other time we humans were in the vicinity, we urinated on the perimeter fence. This proved an excellent deterrent—yet another example of how small sanctuaries must think outside of the box!

We continued to name Momma Memphis's brood after influential activists working for peace and love: Buddha, Corrigan, Dalai, Gandhi, Jane Goodall, Nelson, and Anita. That last—an orangey piglet with black spots—was named for Anita Krajnc, the founder of the Animal Save Movement, who is also a contributor to this book. All of Momma

Memphis's new family were healthy, and soon all seven were running around, as carefree as only piglets can be.

After our first year in Tennessee, I knew that the Pig Preserve model was working. As our reputation grew, we began to receive inquiries from researchers, animal activists, and other sanctuary owners from as far away as Australia, South America, and Europe.

At night, when I slipped into one of the big barns to check on our pigs and heard peaceful snoring, broken only by the occasional grumble as two pigs discussed sleeping arrangements, I felt content that what we were doing meant something. When I looked out over a pasture and saw a couple dozen pigs quietly grazing, I experienced the serenity that comes from knowing that every animal in that field would have been dead were it not for our preserve. Occasionally, after a difficult day, when I was numb with exhaustion, I would lose that sense of self-fulfillment and I had to search for it again. Then, a new pig would arrive, step out of the trailer, experience sunlight, green grass, and blue skies for the first time, and take a few tentative steps of freedom, and my serenity would return.

I could finally and honestly say: *"Yes, our pigs are happy."*

11

On Their Own Terms

JUST AS AT MINI-PIGS, LAURA and I continued to pay all the expenses at The Pig Preserve, except for those for the direct care of the pigs. Since much of our donor base in Virginia had been tied to individual miniatures, we gave up those funds when we transferred those pigs to other sanctuaries. Now, in asking people to sponsor a pig, we continued to tell them that it costs about $50 a month to maintain a potbelly and about $100 a month to maintain a farmed pig. I didn't know how precise those benchmarks were, but certainly each pig we took in increased our bills by hundreds of dollars a year.

We always had a monthly vet bill because we always had animals on medication, especially our elderly or compromised pigs. If we had to send a pig to the University of Tennessee College of Veterinary Medicine, that might cost anywhere from $10,000 to $20,000. For these emergencies, paid ads generated a good response. Otherwise, they weren't worthwhile.

Fundraising remained the one nagging fear that could give me sleepless nights. A sanctuary is a labor of love that must be run like a business. At the prompting of several savvy activists, The Pig Preserve put up a Facebook page, and I plugged and chugged along on social media. Probably, the growing reputation of the sanctuary did more to bring in donors than anything I could do as a low-tech guy in a high-tech world. Between the money raised and our personal subsidization, we still usually ended each year a few thousand dollars in the black.

A year after settling in Tennessee, we created The Pig Preserve Association, Inc. (TPPA), a nonprofit organization fed by private donations, to provide assistance to pig owners for medical and other emergencies and to sanctuaries facing disruptions or bankruptcy. When raising money for both TPPA and The Pig Preserve became too confusing for us and for our donors, we dropped out of TPPA, but, as it happened, not before an adventure or two under the TPPA banner.

In the spring of 2007, a story broke that members of the sheriff's department in Tularosa, New Mexico, were shooting pigs at the home of a member of the local community. This alarming news spread at lightning speed via a flurry of emails and on social media platforms, accompanied by photographs of dead pigs. The response zigzagged between tears and outrage.

A call to the local sheriff's department in Otero County led me to suspect that rumors might have inflamed an already bad situation. I communicated with the rescue groups that had requested help, along with sanctuaries that might be involved in a solution. Then, I flew to El Paso and drove a rental car to the rural area of New Mexico where the animals were located.

Nothing beats an on-site investigation.

The short, verifiable version of the story went like this: A large group of pigs—potbelly–feral mixes of varying ages—was living in woefully substandard conditions in a private home several miles outside of rural Tularosa. The guardian, an elderly lady, could no longer adequately care for her herd or keep the pigs contained on her property. The emaciated pigs—intact boars, sows with nursing babies, and a large number of piglets—were relatively free to roam the rural desert in search of food and water.

Over many months, in the pigs' attempt to survive, they became a nuisance to neighboring farms, destroying alfalfa fields, seeking entry to horse-feed barns, and threatening those who tried to drive them off. Numerous futile attempts had been made by the local sheriff's department, animal control, social services, mental health professionals, and environmental and zoning experts to help the guardian deal with her escalating problems. As the situation grew more unruly, unverified reports circulated that the pigs were now biting people and horses.

On the day of the incident that functioned as a lightning rod, the sheriff's department had responded to yet another complaint by neighbors. This time, when the guardian was confronted, she authorized the sheriff's department to take control of her pigs. This included killing them. As widely reported, several pigs were shot by sheriff's deputies on the grounds that they had behaved with aggression.

My guess was that the pigs, cornered by the deputies, had responded out of fear or in defense of their young, and the deputies took what they considered appropriate action. In fairness, if they had chosen to, they could have justified shooting every pig on the premises.

Several pig sanctuaries that were intimately involved in this situation asked the sheriff's department to be allowed to save the remaining pigs. We were given not only permission but full cooperation from all of the local authorities. For temporary on-site relief, we also received the farm owner's permission to feed and water the pigs; to rebuild their pens; to assist a veterinarian in capturing the pigs to be neutered, blood-tested, and microchipped. All the pigs were then transported to a temporary holding facility at the local vet's clinic, the bulk being then transferred to Ironwood Sanctuary in Arizona, where they would stay until permanent homes could be found.

Though several pigs had sadly lost their lives, we could find no villain. It was obvious to those of us who dealt with the pigs' guardian that she loved her pigs, though she was incapable of looking after them. The Otero County and Tularosa officials behaved professionally in a complex, deteriorating situation. TPPA's goal was to save as many pigs as possible, and that was accomplished. Total rescues: ten mature pigs and twelve piglets, many an attractive mix of potbelly and feral, unsocialized but full of potential.

Though it was satisfying to participate in the New Mexico rescue, it was a relief to get back to The Pig Preserve's peaceful environment. One of the many beauties of our Tennessee property is its free-flowing springs and ponds. Our pigs were seldom more than a hundred yards from a natural water source, which meant I was spared the endless filling of troughs for everyone except our elders, located up at our main barn. I didn't want an arthritic old pig like Rufus or Rosebud to have to choose not to have a drink if that meant struggling up and down a hill to reach the closest pond. It was so much easier for me to put a trough of fresh water by their barn on a daily basis.

In the summer, the sanctuary's feed bill would be $2,200 a month for approximately one hundred pigs. In winter, it could easily double. At our peak population of 175 pigs, it averaged $1,200 *each week*. This did not include the cost of apples (a bushel a day) or of the special diets and daily medicine for our elderly and special-needs pigs. A new barn or shed might cost anywhere from $3,000 to over $4,000. Larger barns, like the twelve-by-thirty-six–foot ones, frequently cost up to $10,000. Each week in winter, we went through dozens of bales of hay to provide fresh bedding in our twenty-four barns. Now that we were cutting from our own two pastures, we could sometimes make it through winter without having to buy more. When defeated by the weather, we had to scramble to find a local source, often at outrageously high prices.

Much of the difference between our summer and winter feed bills resulted from our pigs' seasonal opportunities to forage as they would in the wild. Despite my many years of working with pigs, I had never considered them grazing animals like horses and cows. In Virginia, we had had too much traffic across our pastures to grow much grass, though we had tried—valiantly. Now, with verdant pastures, our pigs could graze from sunup to sundown. They also cleared all the underbrush from our woods. Previously, it had been like walking through barbed wire with its creepers and thorns. Now, it was like a park. They ate the leaves from saplings and stripped the bark from certain trees, especially the pines, which they treated like after-dinner mints. They ate blackberries.

They ate grubs. They ate nuts, foraging right alongside our deer herd and our flock of wild turkeys. When the grass turned brown with Tennessee's first hard freeze, they flipped it over and ate the roots.

From late spring till early fall, we fed our healthy adult pigs only in the evening, as a supplement to what they were harvesting from nature. Though I had grown used to the fact that the gift of freedom meant our pigs would spend the bulk of each day in the company of other pigs, I still made a point of examining and touching each and every one to make sure they were healthy. That was why I fed each social group separately, and why feeding took between three and four hours.

Once cold set in, twice-a-day food became important in order to provide the calories and nutrition our pigs needed to survive and thrive. Much had changed since those early days when I first opened a tin of dog food for our Patti. For many years, we fed Patti and our other potbellies miniature pig food made by Mazuri, a subsidiary of Purina. When our growing herd made that prohibitively expensive, I consulted with veterinarians and the swine nutritionists at the Culpeper Farmers' Co-op to create a mixture designed for optimum pig health and longevity. We created a pellet that combined fat, protein, minerals, and trace elements. Later, when some of my elder pigs developed ulcers, I suspected the corrosive nature of the pellets. In Tennessee, I asked Top Milling to create an all-grain feed in pellet form. We came up with a mixture of cracked corn, wheat, oats, barley, rolled oats, and soybeans, combined with a binder of molasses and pure vegetable oil. The result was a very palatable feed with no additives and no preservatives. It looked and smelled like a granola bar.

Since Top Milling's operator was himself a farmer, he suggested having the soil from each of our pastures analyzed. When those results came back, we discovered the earth where our pigs foraged was well balanced, except that we should add some salt and some calcium carbonate to their feed. We also added diatomaceous earth, a natural wormer, to help keep our pigs parasite-free without using harsh chemicals.

When we moved to The Pig Preserve, I hadn't sufficiently taken into account the difference between operating a one-hundred-acre sanctuary as opposed to a seventeen-acre sanctuary. With everything so spread out, maintenance that could have been handled with a short walk now required tools and supplies to be driven to the site. I also had to devise a repertoire of options for lifting and transporting a sick or injured pig from anywhere on the property to the main barn for continued care. Though I became adept at using ropes, pulleys, slings, and other innovative tools to manage on my own, I finally had to concede the need for extra help.

While consulting with the local mill operator about the preparation of our feed, I met one of the employees, Tim Lewallen. In 2017, when Top Milling closed down, we hired Tim full-time. It was a good call. While my body was still strong, it was an aging body, and Tim had a solid work ethic.

The addition of another person proved especially useful when we doled out that carefully mixed grain to our herd. Until a couple of years ago, feeding the pigs by social group at the troughs located by their sleeping barns had become a fairly routine one-person operation. I would drive our tractor to the eight or nine stations in the upper part of the sanctuary, pour the feed into the troughs, then leave the pigs to jostle for place while I headed to the bottom half of the sanctuary.

All this civility broke down when we took in sixty-six young pigs within a year, some as rescues from hurricanes that hit the North and South Carolinas. Our population, which had been relatively mature, changed dynamically with the addition of these energetic risk-takers; it was like adding sixty-six teenagers to a nursing home. Instead of simply eating at their feeding stations, these youngsters were smart and ambitious enough to know that if they followed the tractor, they could snatch more food at every other station.

Feeding turned into a free-for-all.

Either Tim or I would drive the tractor. The other, riding on the back, would jump off at each station to fill each trough, while dodging anywhere from twenty-five to forty juveniles running from trough to trough to trough, trying to snatch every last bite. Of course, the elders resented this intrusion. Using their big, slow-moving bodies and squeals of protest, they would drive off the youngsters.

I admit that it was fun to watch—total mayhem! But it was also dangerous. If Tim or I were caught between a pig and a food trough, that pig would draw a direct line through us, knocking us out of the way as they would another pig. Farmed pigs don't understand how big they are in comparison to humans. They don't consider that they have a nine-foot body behind their ears. Meanwhile, the lighter pigs make up for their lack of heft with energy. Managing that crowd required a football running back's power combined with a ballerina's agility, and I was neither as fast or as graceful as I used to be. When it was muddy, keeping two feet on the ground was especially tricky.

These skirmishes served as a reminder of the dangers that lurked every day when dealing with large animals and heavy farm equipment. What is natural for a 1,200-pound pig competing with a couple of 450-pound potbellies is not natural for a 250-pound man who becomes caught among them. While working alone, which I have done 95 percent of the time, I have been knocked out on at least four occasions. Once, I almost froze to death. It takes only a momentary lapse in concentration or good judgement for some routine task involving a sharp-bladed power tool to turn into an emergency. With only sporadic cellphone service, my list of mistakes, miscalculations, and just plain stupid moves has led to an impressive assortment of injuries over the years.

Because The Pig Preserve is on the slopes of the Cumberland Mountains, we didn't have summer's brutal heat and our winters were mild, with perhaps ten days of snow that soon melted. Digging pigs and barns out of icy drifts in a raging blizzard became a fast-fading memory. Though once-a-day feeding cut down on my summer chores, our pigs' foraging created another sort of labor. Pigs have a genetic imprint measured in acres. It didn't take long for 175 of them to exhaust what was naturally available on one hundred acres. In the wild, the herd would have moved to another forage area. At The Pig Preserve, they were constrained by fences. I used to think that one acre could support four pigs. I have now come to believe that it's more like one acre for two pigs.

In the spring, I had to do considerable repair work on the pastures. I'd drive the tractor to areas where the pigs had done most of their rooting, smooth the earth out, then sow five hundred pounds of grass seeds for future grazing. I'd plant at least two dozen saplings, specifically chosen to provide forage. One year it was apple trees, the next it was figs, then it was nut trees.

Running The Pig Preserve was like working a small farm. It was mostly blue-collar, dirty, mundane labor. Much of it was mind-numbingly repetitive, with the pigs always wanting to help with projects. No matter what I was doing, I could always count on a dozen or so pigs of varying sizes stopping by to offer their assistance by stealing tools, knocking over ladders, tearing open bags of cement, upsetting buckets of nails, and scratching their backs on a post that I had spent an hour setting plumb. A two-hour job would take three or four, but I couldn't stay angry with them. It was their nature to be inquisitive and to find whatever I was doing fascinating.

Our house, along with our acreage, slowly evolved to suit the needs of the pigs. Our two bedrooms in the basement, where the original owners had first lived, became a storage facility for medical equipment, fans, hot-water tanks, and so forth. The basement living room and kitchen, with their concrete floors and air-conditioning and heating systems, became a pen for newborn piglets, pregnant sows, sick animals, and those recovering from surgery. A spiral staircase from our living room down to our basement allowed us to monitor its residents 24/7. A sizable entrance at the back of the house made for easy, daily cleaning. It wasn't a pretty place, but it was functional.

Even our upstairs living quarters were often placed in service as a nursery or temporary home for special-needs pigs. Four miniatures—Butch, Sundance, Toot, and Puddles—spent their first few months there waiting till spring, when they could safely be put out with the larger pigs.

While they were still in residency, we welcomed Jasmine, a baby farmed pig who came to us from Franklin, Tennessee.

A young couple were out driving when they saw a mass fall out of the back of a pickup truck, then roll to the side of the road. On stopping to check, they found a little pig covered in road rash. After locating our sanctuary on Google, they called and described their rescue as "almost unconscious and bleeding everywhere." I asked them to take the pig to the nearest vet to see if she was salvageable. In many cases, a small pig who has fallen from a truck will have so many injuries that the most humane response would be to euthanize her.

The vet reported that while the pig looked a mess, he could see nothing wrong with this little female beside scrapes and road rash. After he gave her some formula, the couple wrapped her in a blanket, then brought her from Franklin to us, which was about a four-and-a-half–hour drive.

Jasmine arrived at the sanctuary just before dark. I rushed her to our local vet, who confirmed that she was fine except for the sound of fluid in her lungs. To prevent this from turning into pneumonia, she was given antibiotics.

With her strong will to live and marvelously loving personality, Jasmine became one my favorite, best-beloved pigs. She was also our fifth in-house baby, along with eight dogs and God knows how many cats. Yes, it was chaos for a while, but Jasmine made a quick recovery, then found her place in the herd. So did Butch, Sundance, Toot, and Puddles, quickly earning the nickname "The Demons" for their rambunctious, high-spirited behavior.

The following winter, our living quarters were occupied by Finnegan, a young Yorkshire pig born with no front feet. Finnegan took immediate possession of our couch, where he could observe the dogs and cats from his lofty perch without having to tangle with them. Several of the dogs took to sleeping on the couch with Finnegan, so by winter's end, it was suitable only for a trip to the dump. Laura and I had long ago realized that if you run a small sanctuary, your house will never be a candidate for an article in *Better Homes and Gardens*. Those who understood our mission would be okay with that. The others didn't matter.

As winter turned into spring, we were troubled about what to do with Finnegan. As a Yorkshire, he would soon be too big to stay in the house, yet he was too compromised to be placed with the rest of the herd. By taking him into the yard a few times in early spring, I discovered that in soft grass and dirt, he could move around without too much discomfort, even with missing feet. Now he was eager to get outdoors and to become a pig . . . but a solitary pig?

As so often happened in the life of our sanctuary, fate provided a solution.

Yet another small farmed pig had fallen off a truck on his way to auction. Though the local sheriff's office had been called to euthanize him, the responding deputy didn't have the heart to shoot him. A local dog-and-cat–rescue group claimed the little pig, then called me. After driving to Kentucky in a thunderous rain storm, I retrieved the little pig, then took him directly to the University of Tennessee College of Veterinary Medicine, arriving at around 11 p.m.

When I checked our rescue into UT, the on-duty vet asked his name. I replied, "He fell off a truck on his way to an auction, so let's call him Lucky."

Lucky had suffered a badly fractured left front leg and shoulder, which Dr. David Anderson called "catastrophic." As he cautioned, "If we can't provide him with four legs, he's going to have a very difficult time."

The commonsense response would probably have been to put Lucky down, but I wanted to give him a chance at having a life. While Lucky's surgery was successful as far as it went, he had nerve and muscle damage, resulting in a limited range of motion in his leg. For a Hampshire who could grow to 1,200 pounds, that was troubling.

Because I worried about Lucky's limited use of his recovering leg, we returned him to UT for another surgery. This time, Dr. Anderson released some of the pressure in his leg in order to straighten it, hoping to give Lucky more mobility. This worked well enough for a while, but as Lucky grew heavier, a third surgery was required, in which Dr. Anderson fused his joints, leaving him with a stiff leg he could use as a crutch.

As Dr. Anderson commented after getting to know Lucky: "He's unique. To quote a spider in a web, 'That's some pig!'"

I agreed. Lucky had a great love of life. He wanted so badly to be like other pigs. How could we make this happen? Given the fragility of his leg, I could not allow him to run free with our other farmed pigs in case he got into a fight or tumbled down one of our rocky bluffs.

A plan began to emerge: I had planted a huge garden of corn, tomatoes, melons, squash, and cucumbers, for us and for the pigs. I had even installed a tiki bar with thatch roof, pastel beach chairs, a fire pit, torches, and a sign that said MARGARITAVILLE because Laura was a big Jimmy Buffett fan.

After I had invested a lot of effort in building our tiki bar, Laura and I lacked sufficient leisure time to put it to good use. In fact, in the four or five years of its existence, I believe we used it once. Now, I looked at our tiki bar, and I said to Laura: "It has fencing on two sides. It would make a nice pen for two pigs."

I plowed up our vegetable garden and knocked down Margaritaville, then put in a barn and fencing and gates. Laura hated the destruction of Margaritaville, and I mourned the loss of the garden, but the pasture was easily visible from the house so we could monitor it. It had a large number of shade trees, a huge grape arbor, and a wisteria vine providing shade from the summer heat—ideal for two compromised pigs, provided Finnegan and Lucky could tolerate each other. Although Lucky, at 250 pounds, was twice Finnegan's size, they had similar dispositions. Both were energetic, mischievous guys in spite of their handicaps. Something told me that they would do well together.

When Lucky came home from his last surgery, we set him in this new pasture with its nice, soft grass. We left him by himself for several weeks so he could acclimate to living outdoors after having spent his entire life either in the confines of a factory farm or in a pen at the University of Tennessee. Once Lucky seemed comfortable in his new home, we introduced Finnegan—with a great deal of trepidation.

Their meeting turned out to be anticlimactic in a good way. Lucky immediately made it clear to Finnegan that he was the "senior pig," and Finnegan apparently couldn't have cared less. He was content to follow Lucky around, probably just super glad to have the company of another pig. The two slept together that night. Within a week, they were inseparable.

During the next several years, Lucky and Finnegan's one-acre pasture became an ideal place for newly arrived piglets, along with others still too small to be introduced to our mature herds. Lucky and Finnegan seemed to enjoy hosting the babies, and the babies were agile enough to playfully evade Lucky and Finnegan when they tried to "boss" them around. At night, they snuggled under the big pigs for warmth and comfort.

Lucky and Finnegan lived happily together for more than eight years. Then, we found Finnegan dead in his pasture one morning. His unexplained passing was a shock to us. Whatever birth defect had caused him to be born without feet had likely caused other issues that weren't readily apparent. His necropsy at UT was unable to pinpoint anything specific but found a number of cardiac abnormalities and other internal concerns.

A few months after Finnegan's death, Lucky began to have difficulty getting up and walking. Soon he quit eating. In what was becoming a familiar pattern, Lucky never recovered from the loss of his pal, though we tried everything we could to pique his interest in life. A large, elderly pig with mobility issues goes downhill quickly. Their muscles atrophy, and their unused leg joints lock up, then fail. Finally, we took the difficult decision to euthanize Lucky. His sad denouement after Finnegan's death was yet another demonstration of the depth of pig emotions and the power of pig grief.

One of our primary reasons for moving to Tennessee had been to increase our commitment to farmed pigs like Lucky. Days of human celebration were especially deadly for them. One year, in the approach to Easter, a friend of ours noticed a post on Craigslist: "Barbecue pig, 6 months, 150 pounds. Your Easter feast. $80.00." An accompanying photo showed a forlorn little red pig living in a horse trailer. Apparently, the seller had intended to barbecue her for his own Easter dinner but decided instead to flip her for money.

When a sanctuary friend contacted us, we agreed to take this little girl, but there was a catch: As a nonprofit corporation, we had a firm policy of not buying animals. This is a common rule among animal sanctuaries, and one grounded in good sense. Buying animals to keep them from slaughter is not only a fast track to bankruptcy, but it also encourages opportunists to breed them for sale.

After our friend had raised the money to purchase the pig, Laura and I were happy to make the two-hour drive to pick her up. Our little redhead, now named Easter, was anxious at first. Once she seemed to realize her life was about to change for the better, she hopped in the car, and off we went.

We treated Easter for hog lice, then invited her to enjoy Easter dinner with us—the vegan way of having a pig for dinner. Our fiery redhead turned out to be a social butterfly who integrated seamlessly into the herd, then became a great hostess to newcomers. It was the kind of easy transition that creates a wave of elation, sweeping away any sense of sacrifice and assuring us that our work was worthwhile.

Abbey Christopher was listed online as "a Christmas ham." Using contacts in Arkansas, a friend in California purchased this young pig. Since the rescuers didn't even know the gender, they chose "Abbey" should the pig prove to be female and "Christopher" should they be a male. Their pig was definitely a male—a handsome black Berkshire with the classic four white socks and a white blaze on his snout.

Abbey Christopher, who now weighs well over one thousand pounds, so thoroughly enjoys the companionship of humans that he rubs against us like a dog, nearly knocking us over with his enthusiasm. He is a big, slow-moving giant who gets along with every other creature on the property, and who seems to wear a perpetual smile on his huge face.

Bubby and Sissy were brother and sister Hampshires, also weighing about one thousand pounds each. As babies, they were purchased by a homesteading family to be raised for meat. After the family fell in love with these young pigs, they just couldn't go through with slaughtering them. Though the man of the house was adamant that they could not keep them as companions, he did agree on a reprieve. In desperation, the mother began a computer search to find options for these two rapidly growing youngsters.

As so often happens, her search led to our sanctuary. Although we were slightly above our self-imposed limit, she pleaded so desperately about the fast-approaching end of the grace period that my heart overrode my brain and we agreed to accept the siblings.

Just one problem: Bubby and Sissy lived in Seattle, Washington, and their guardians were too poor to transport them to us in Tennessee. This is not unusual. All too often, it's the logistics of transport that sink a rescue.

In need of a small miracle, I phoned my friends, Bill and Marie Worley, who had an animal-transporting business. I was hoping that their schedule, based on contracted loads and pre-determined delivery destinations, might accommodate Bubby and Sissy. Amazingly, they were planning to be in the Seattle area before that ominous deadline, and yes, they would have room in their trailer for two small pigs.

Still, there was another hang-up: The Worleys, who operated out of Texas, had been gifted a Christmas cruise to Mexico by their daughter, which meant they could not deliver the pigs directly from Washington to Tennessee.

Fortunately for Bubby and Sissy, the Worleys are big-hearted "pig people." They offered to pick up the siblings in Washington before their deadline, then transport them to their Texas ranch, where their daughter would care for them until after their cruise.

So that's what happened. On returning from vacation, the Worleys loaded up the two young pigs and transported them to our sanctuary.

So many Good Samaritans. Two more lives saved!

Shortly after we moved to The Pig Preserve, an event occurred that convinced both Laura and me that we had been right to discontinue our adoption program.

At Mini-Pigs, we had rescued Peaches from a trailer park in Virginia, then adopted her out to a family who had a potbelly named Daisy. Because the family's twenty-five–acre farm was near us, we were able to watch Peaches develop a close bond with Daisy, and then grow into a pretty, sociable little pig.

This adoption seemed ideal.

We lost touch with this family when we moved to Tennessee, so we were taken by surprise when they returned Peaches to us. She had a medical problem their local vet could not diagnose. They also surrendered Daisy on the grounds that Daisy and Peaches were inseparable. Peaches was thin, pale, and suffering from constant diarrhea. After some difficulty, we received a diagnosis of stomach ulcers. When properly medicated, Peaches dramatically improved, but she continued to have recurring bouts.

We kept Peaches as a yard pig to closely monitor and medicate her while settling Daisy in one of our pastures, where she easily found a family group. As we tried different treatments, Peaches' condition would improve for a while, only to deteriorate once again.

Despite her health problems, Peaches remained a happy little pig. Since she loved being around people, she thrived on the special attention we gave her. She put on a proper amount of weight and enjoyed the companionship of the other yard pigs, even our dogs. Peaches' days were spent lying in the sun, receiving belly rubs, and tending to her special relationship with Wilbur, one of our elder potbellies. He and his sister Henrietta had been rescued as piglets from a "farm-to-table" exhibit at one of the historical museum farms.

When Wilbur wasn't feeling well, Peaches was by his side. When Peaches wasn't feeling well, Wilbur let her move into his doghouse, despite the unpleasantness of her condition.

Eventually, Peaches' diarrhea would not go away. For the first time since her treatment for ulcers began, Peaches quit eating, despite our coaxing. As she grew weaker, we moved her into the house and took turns staying with her. When it became obvious that Peaches had given up, we made arrangements to euthanize her. Two days before her scheduled appointment, Peaches died peacefully in Laura's arms.

We were sad but comfortable with her death. The damage done to her system by the long period without treatment for her ulcers had proved irreversible. Peaches bore her affliction quietly and with patience. She was never cross, and never angry. Her courage in dealing with her illness was inspirational, till finally she reached the point where she was tired of being sick. She was tired of hurting. It was our responsibility as caretakers to allow her to go peacefully and with dignity. She crossed the Rainbow Bridge knowing that she was loved and that it was okay for her to leave us.

Peaches' death left a hole in our sanctuary family. The grief Laura and I always felt when holding the head of a sick, dying pig is impossible to describe. Our minds are fully prepared for the eventuality of the death of every animal in our care, but our hearts are never ready to let one of our dear friends go.

12

Talking with Pigs

While raising Patti Murphy, we learned that pigs were very vocal animals who weren't shy about expressing their views. Though animal scientists may disagree, I believe pigs have about forty or more distinct vocalizations, which can be broken down into grunts, squeals, barks, and oofs. They combine these with twenty to thirty different body movements to create close to one hundred distinct expressions. A pig's squeal can signify anything from delight to terror. That dramatic range is also true with grunts. Context and body language are everything when interpreting these expressions.

When pigs meet, whether or not they're from the same social group, they go snout to snout and grunt deeply at one another for about thirty seconds. Then they separate and move on.

As pigs meander slowly down to a pond for a warm afternoon swim and mud bath, they utter the soft, deep, guttural grunts and oofs of contentment. And yes, I am convinced pigs talk to themselves. Frequently, I have observed a pig walking by themselves while uttering a whole plethora of low-pitched grunts and other vocalizations, like a muttering little old person. Often, they even act embarrassed when they discover I am listening and chuckling.

Sleeping pigs frequently give off a series of mumbled, soft grunts and high-pitched squeals, in addition to their inveterate snoring. When a pig is receiving a belly rub, they make a very deep rumbling like that of a Harley Davidson motorcycle idling at curbside.

Pigs also have a unique sound for grieving over a dead beloved pig. It's a soft "whooping" made with their mouths open, almost like mouth-breathing, that I've never heard them use in any other context.

At feeding time, I am very familiar with our pigs' whines, squeals, and barks of impatience. I could look across our property at two thirty on a hot or cold afternoon and not see a pig because all are wallowing in their mud holes or snuggled in their warm barns. Then, at three o'clock, every pig will be at their feeding station. If I am late, they squall—whine, squeal, and bark—rising in a crescendo.

Sir Winston Churchill once said: "I am fond of pigs. Dogs look up to us. Cats look down on us. Pigs treat us as equals."

Churchill was right. Pigs will carry on a conversation with us, fully expecting us to understand what they're saying. To work with a herd of pigs, I've had to maintain myself as the "alpha" to prevent situations from becoming chaotic and to keep myself from being bullied by the more aggressive pigs. As I learned from Patti, I have to do this in the same way a more dominant pig would correct a subordinate. When a dominant pig is challenged, they emit a series of grunts and squeals, which may become more insistent depending on how quickly the less dominant pig obeys, and perhaps end with a swipe of the head as an exclamation mark. Usually, all I have to do is raise my voice (my "grunt") or give a pig a shove on the flanks to correct their behavior. Once in a while, a pig (almost always a miniature) will challenge me. I know I must not back down or a shoving match might ensue. Eventually, the challenging pig retreats, though not without a great deal of grumbling and grumping.

Sometimes, I have been bitten or, more frequently, "tusked" by a male pig. These injuries are not intentionally caused. It's easy to get hurt when crowded by several rambunctious pigs, or when trying to coax a sick or injured pig into a stall or trailer. Bumps, bruises, pratfalls, and the occasional tusk slash are all byproducts of a pig's physical vocabulary, unintended or not.

All pigs have tusks, which are their upper and lower canine teeth. A male's tusks continue to grow throughout his life due to testosterone. This happens even after he's been neutered, but at a slower pace. We don't typically cut tusks unless they grow long enough to injure other pigs during play or hurt us when we work with the pig, or unless they are likely to grow back into the pig's face. Most of the time, through normal rooting and playing, and just being pigs, males wear down their tusks or break and blunt their tips.

Occasionally, when a pig's tusks aren't quite ready to trim, that pig will catch one in a fence. The pig's natural reaction to any fearful stimulus is to withdraw, so he pulls back, hooking himself into the fence even more securely. Worse, if he steps forward, he might catch his body. Then, when he pulls back, he is likely to cut himself. On realizing he is trapped, he starts this god-awful, clearly recognizable scream of terror. Every pig within hearing distance immediately comes running with the idea of helping. They don't know how to help, so they stand around screaming with him. The more frantic the hooked pig, the more frantic the screaming chorus.

As soon as I hear that first distinctive, terrified scream, I know one of our males has caught his tusks, most likely in a fence. My job is to rush to release him. That is why every tractor or four-wheeler or other piece of equipment on our property carries its own set of bolt cutters. To rescue the hooked pig, I sometimes must wade through twenty to thirty crazed pig supporters, knowing that in their panic, they might think I am trying to hurt the trapped pig and might attack me. This is one of the more dangerous situations that can crop up anywhere on the property. I always make it a practice to cut the fence so the pig can untangle himself. I never cut the tusks. Afterwards, it might take the assembled pigs as long as half an hour to calm down and return to their normal, complacent selves.

Pigs have a herd signal for danger, which is also a unique and easily recognizable series of sharp barks. Remember: these are prey animals, regardless of their size. Within seconds of that sound, every pig within hearing range will instantly turn and run blindly while taking up that bark, resulting in a deafening cacophony. Occasionally, I have sheltered in the house during a driving storm, watching thirty pigs blissfully graze in a pasture, with lightning flashing and thunder crashing. Three days later, I might be out in a pasture with that same group on a sunny day, when a dust devil or a swirl of leaves would kick up, or maybe something would fall from a tree. One pig suddenly raised her head, gave the danger grunt, and thirty pigs stampeded in terror. Apparently, if you're a pig, you innately know that if you stop to wonder, *What's this all about?* you might be a predator's next meal.

This reflex action is another concern for anyone working in the field with pigs, because you never know what's going to trigger that response. Whenever I hear that sound, I race in the safest direction. Otherwise, I might get trampled by one or more 1,200-pound pigs traveling at twelve to fifteen miles per hour. Farmed pigs aren't long-distance runners, but once they power up for a spurt, it takes a while for them to brake, especially if they're running downhill and it's muddy. By the time they hit the middle of the hill, they're in full, breakneck stride. In more than a few cases, we've had pigs whose brakes failed crash into a barn or take out a large farm gate.

If I am in a safe place when I hear a pig's distinctive bark of danger, I check to see what caused it, in case it's a predator. Many wild animals roam free in this remote area of Tennessee. We have extended the concept of sanctuary to all of them, with one ironclad rule: *Whoever you are, you're welcome, so long as you can get along with everybody else.*

We have a substantial population of black bears who regularly roam the sanctuary. They have never bothered a pig, and when I have encountered one in the woods, it has always been a case of mutual surprise. I see the bear at about the same time as the bear sees me. He stands up on his hind legs. I stand in place, facing him. We eyeball each other for about thirty seconds. Then, the bear turns, drops down on all fours, and walks calmly off into the woods. Once I had identified a frequent bear crossing area, I cut down the fence to allow the bears free movement. That means they don't have to climb over my fences, and I don't have to constantly repair them. A win–win situation.

When Laura and I first arrived in Tennessee, we had a significant coyote problem. We didn't want to go with trained guard dogs or other guard animals such as donkeys or llamas. Instead, the sanctuary has always supported a rescue population of around eight to ten dogs, which is sufficient to keep the coyotes at a distance. Though coyotes aren't going to mess with a large farmed pig, a pack might go after a potbelly or a young feral. Whenever coyotes evade our dogs and come a little too close to our pigs, I shoot over their heads with a rifle, or into the ground in front of them to spray them with dirt. That noise is enough to drive them off, so I've never had to kill one. The days are long gone when I hunted in Panama or as a young Marine. After my overseas tour, I completely lost interest in killing.

We once had a fairly large population of Copperheads and Eastern Diamondback rattlesnakes, but the vast majority of those moved on. I think snakes work off of vibrations, and a bunch of 1,200-pound animals tramping through the woods may have convinced them to relocate to a more peaceful neighborhood. Pigs are also adept at killing snakes. They stomp on them with their hooves.

Copperheads are relatively small with fairly short fangs. Their venom is usually injected subcutaneously, creating a localized injury rather than a systemic poisoning. We treat a pig who has been bitten for swelling, nausea, and fever and with antibiotics to keep infection at bay. Often, the pig sloughs a lot of skin around the bite, so we must keep the area clean and bacteria-free while new skin forms.

Eastern Diamondback rattlesnakes can grow to over six feet. Their bite is much more serious but much less common. With their very long fangs, they deliver poison deep into muscle tissue. For smaller pigs, this can be fatal. For larger pigs, it depends on the location of the bite and the amount of poison injected.

Whenever I found a poisonous snake in one of our barns or where our pigs congregated, I killed the snake. I rarely had to do that because of our sizable population of black snakes, which we encourage. Black snakes are immune to a Copperhead's poison, so they kill and eat Copperheads. If I have a black snake living in each of our barns, I don't have to worry about Copperheads. Black snakes also help with rodent control.

We have snapping turtles in one pond who are not big enough to aggravate or threaten our pigs. Unfortunately, our population of owls and hawks has taken a substantial toll on our cats who ventured out into the pastures. Usually, our cats prefer to hang out in the barns. We anchor food bowls for them up in the rafters, where the pigs can't get to them. Usually, our cats don't need that additional feed. Helping to keep down the rodent and snake population makes them self-sufficient.

We've never had a big cat problem. The preserve is close to the Daniel Boone National Forest and Big South Fork, which is a gigantic natural wildlife recreation area. This means panthers and cougars have thousands of acres of pristine habitat where they can be themselves and don't need our property.

We have healthy populations of deer, squirrels, and rabbits. We also have about sixty wild turkeys, and fishes in our ponds. We have never allowed hunting or fishing. We're surrounded by hundreds of acres of deep woods, with well-maintained fences and no back roads. Hunters have many places better suited to their needs, with easier access. Most of them also know that if they shoot too close to the preserve, I will probably shoot back.

All in all, I consider pigs to be among the most honest, most communicative creatures on Earth. A pig who is angry, scared, or happy will let you know in no uncertain terms. I have found only one exception to this rule of forthright self-expression. When a pig is sick, injured, or in pain, they become the most stoic of creatures. I believe this goes back to their being a prey animal. When a herd is being stalked by a predator, a pig who is weak or obviously compromised becomes the most likely candidate for attack. Hiding their vulnerability so as not to draw attention to themselves is a pig's survival mechanism.

In a domestic situation, a pig's stoicism becomes a liability because it delays treatment. Usually, our first clue that an animal is becoming sick or has been injured is that they begin to act "differently." Maybe they will hang back at feeding time. Maybe they will spend more time alone. To spot these subtle differences, a good caretaker must learn the personality of each pig and maintain a comfortable relationship with them. That is easier with some pigs than with others, especially since we at the preserve have a long history of taking in hard-to-handle pigs. Even these so-called "problem pigs" will usually settle in within thirty days or so, not because I'm a freaking pig whisperer but because aggressiveness is just a pig's way of trying to tell you something is wrong in their world. Often, it is something simple. Fix it, and these "difficult" pigs often become the biggest love mooches.

Big Bad Bobby—a large black and white Hampshire—had been raised in a pen by two old Arkansas rednecks. Early in his life, Bobby realized that his size afforded him a measure of intimidation. As an obnoxious teenager, he would throw around his head and snap at his caretakers, demanding more food by scaring the hell out of them. While they loved Bobby, they were, by their own admission, terrified of him.

When Bobby arrived at the sanctuary after a long, tiring trip from Arkansas in a small trailer, he brought his attitude with him. Surly around humans and downright aggressive to all the other pigs, Bobby quickly gained a reputation as a bully. He seemed to delight in challenging any pig who tried to make friends with him and any human who came too close.

Whenever we caretakers were in Bobby's territory, we came to expect he would go out of his way to try to jerk the feed bucket out of our hands, or to snap at us until we dropped the bucket and took off. He ignored voice corrections, and when I shoved him to assert my dominance, he seemed to enjoy it as a game. Finally, in desperation, I started bonking Bobby over the head with an empty feed bucket when he acted out. While this didn't hurt his massive three-hundred–pound head, the noise of the bucket bouncing off his body unnerved him.

I knew Bobby wasn't naturally mean or aggressive. With his four-inch, razor-sharp tusks and gigantic mouth of teeth, he could seriously injure me any time he wanted. With Bobby, I felt sure, it was just attitude. He had probably been a bored and very lonely young pig, confined in his pen. The only way he could convey his frustration was by acting out.

When Bobby wasn't being an ass, I lavished love on him in the form of butt scratches, kind words, and the occasional treat of a large apple. When he was being ornery, he received only harsh words and a substantial "bonk" on the head with the feed bucket.

For the first time in Bobby's life, somebody was standing up to him, and it confused him.

I became totally comfortable in Bobby's presence, and he began to take pleasure in receiving love instead of bonks on the head. As he calmed down enough to enjoy life at the sanctuary, he tried to make friends with some of the other pigs. Unfortunately, he had beaten up or intimidated so many of them that they all ran away, screeching in terror. This seemed to genuinely bother Bobby, but I couldn't intervene. It was a "pig thing," to be worked out by the herd.

After Bobby had been with us for several months, we took in a few good-sized farmed pigs. Since these newcomers didn't know the "old Bobby," they were amenable to forming a friendship with him. One pig in particular, Blue Eyed Wilbur, had been taken from his mother and litter mates at a very young age. Since he had grown up thinking he was the only pig in the world, he felt traumatized by the presence of other animals who looked, acted, and smelled just like him. Even the smallest miniature pigs could cause him to run off, terrified, into the woods.

At some point, Blue Eyed Wilbur and Bobby crossed paths. I will never know what they said to each other at this meeting, because I wasn't present. For reasons unknown, Bobby and Blue Eyed Wilbur became fast friends. At last, Big Bad Bobby had a buddy, and Blue Eyed Wilbur had a partner to stand up for him when the other pigs tried to intimidate him. Both pigs are now among the gentlest, friendliest at the sanctuary. And while not as inseparable as they once were, they generally eat and sleep together. They can also be observed hanging out, side by side, in the woods or down at one of the ponds.

Despite the fact that our farmed and miniature pigs who come to us are the results of generations of breeding in captivity, they still carry all the hereditary skills for seeking safety, for socialization, and for vocalization, skills that were developed when their ancestors roamed free. Humans may have bred farmed pigs for characteristics, such as docility, that make them more easily and efficiently exploited for meat, but pigs have not forgotten how to be pigs. When most step off a trailer into the idyll of The Pig Preserve, they don't need much education. Only a minority have to reach back through years of trauma to make an adjustment before they, too, are oofing, grunting, squealing, and barking as they make friends, settle disputes, warn of dangers, grieve losses, and discover their own special place in the herd.

13

A Pig Is a Horizontal Human

Our pigs possess a psychic gift that seems inborn. Whether we were in Virginia or in Tennessee, each one knew the schedule of our most trusted veterinarian better than we did. The entire herd would be the epitome of health and happiness until our vet decided to take a week's vacation. One day after the vet had left, three pigs would come down with some unidentifiable illness, four would start limping, and a dozen would show up with cuts, gashes, or unspecified abscesses. It was pretty much guaranteed that the vet who was covering for our vet would never have seen a farmed pig since attending veterinary school twenty-five years before and would know nothing at all about potbellies.

When pigs were raised on family farms and in rural households, vets were routinely presented with a great variety of common pig problems. Now that most pigs are raised in factory farms, even rural vets could practice for ten years without ever having to diagnose or treat a sick pig. Researchers who specialize in pigs are usually "experts" on the wrong kind of medical information. All their studies are focused on practices geared towards raising "meat" pigs quickly and cheaply for slaughter at six to eight months. Even information that on the surface seems aimed at improving a pig's health can carry this hidden agenda.

In the early days at Culpeper, we vaccinated all our pigs every year for a host of diseases, including erysipelas, a potentially fatal bacterial infection characterized by fever and paralysis of the rear legs. When I called our vet for Wilbur, who was spiking a fever and having seizures, unable to use his rear legs, we followed our normal diagnostic procedure. That meant making a list of possible causes, from the most likely to the least likely, then narrowing the list down to the most probable cause.

Though erysipelas was at the top of our list based on the presenting signs, we passed right by it because Wilbur had been routinely vaccinated for this particular disease ever since he was born. The more Wilbur was treated for other possible causes, the sicker he became. Finally, I blurted, "Let's treat Wilbur for erysipelas."

That's what we did, and Wilbur was soon cured. That left me with the question: *If Wilbur was vaccinated every year for erysipelas, how did he get erysipelas?*

When I dug into the research, I found that there are twenty-two variations of erysipelas. The vaccine developed for pigs only protects them against two of these. Which two? The two most common in factory farms, where pigs are crowded together in extremely stressful conditions, without access to the out-of-doors. Any concern for the pigs is secondary to meat producers' profit.

Our sanctuary pigs were more vulnerable to some—or all—of the other twenty erysipelas pathogens. In a natural setting, erysipelas lives in the soil and is commonly spread through bird and rodent droppings. Since pigs root, they typically have the bacteria in their bodies. As long as they are healthy, with a good immune system, this doesn't cause them problems. When a pig is stressed, the bacteria may become active.

This information helped me to understand a pattern among our pigs. In Virginia, where very cold days quickly turned into spring's first really hot days, our potbellies frequently fell ill. We thought they were suffering physically from the heat; we didn't understand that the drastic change in weather was itself a stressor, lowering their immune systems and making them susceptible to a host of maladies they otherwise would never have contracted.

The same profit-driven motive that lurks behind the production of the erysipelas vaccine applies to other pig vaccines, which was why we had stopped using them at Mini-Pigs. In the decades since, we have never had a single pig become sick from anything against which we would have vaccinated. Through this experience and many others, I came to the conclusion that pigs have amazingly robust immune systems. If we provide them with a lifestyle that encourages these systems to function at peak capacity, they are healthy. At The Pig Preserve, with fresh air, sunlight, a good grain-based diet, and emotional stability, the number of illnesses dropped dramatically. Our herd became healthier, with many more living into their twenties. And I don't think that's a recipe just for pigs. It works for all organisms, including humans.

Because pigs often came into our care sick or injured, I found my paramedic education a valuable adjunct to the expertise of our vets. During my time studying at George Washington University and in my years hanging around a lot of hospitals as a paramedic, I had also learned that many medical treatments for humans had been developed by research on pigs. This connection was reinforced by pigs who came to us from laboratories where they had been subjected to experiments to benefit humans.

Our Ossabaw Island pigs had been used for human research because inbreeding had resulted in their possessing a low-grade form of diabetes. Bridgette came to us with burn patterns down her back due to studies for burn treatments for humans. We also took in three brothers who had been repeatedly operated on by medical students. Sadly, these brothers, whom we named after The Chipmunks—Alvin, Theodore, and Simon—died at ages four, five, and six due to complications from surgical lesions in their abdomens, which were a direct result of those operations.

My awareness of how often pigs are used as medical stand-ins for humans led me to wonder if medicines and practices developed for humans might apply to pigs. Could there be some payback?

Pigs and humans are both mammals with very similar organs and bodily systems. A pig's respiratory system is almost identical to ours, and they have multi-lobed lungs like we do. A pig's cardiovascular system mirrors ours to the extent that the valves of a pig's heart can be transplanted into a human heart. Porcine and human gastrointestinal systems are basically the same, except pigs have about 30 percent more intestines. That provides greater nutritional absorption, so pigs receive more value from a pound of food than you or I do. Their skin is thicker, and they have a heavier adipose layer, but pigskin can still be used as a graft in human burn cases. Pigs' eyes work the same way as do ours. Unlike many animals familiar to us as house companions, pigs see color just as we do.

I recalled that some of the medical literature that I had studied referred to pigs as "horizontal humans." This led me to ask doctors in the ER where I used to take human patients, "If someone came to you with symptoms A, B, and C, what would you do?" I would then apply their suggested treatment for a human to my pigs with those symptoms.

Over the years, my augmenting of veterinary medicine with human medicine produced surprisingly good results, especially for our elderly pigs. Because factory-farm owners had no interest in raising pigs, except for breeding boars and sows, to maturity, there was a void of medical information on aging pigs. Humans die of cancer. So do pigs. Humans get ulcers. So do pigs. Humans suffer from upper respiratory issues and die from pneumonia. So do pigs. Humans have heart attacks. So do pigs.

Unlike humans, pigs don't drink alcohol, smoke, use drugs, or eat processed food, so pigs raised in a natural environment are probably a whole lot healthier than most humans.

If I consulted a vet about how to treat an elderly, one-thousand–pound pig for congestive heart failure, he wouldn't know how to answer me. If I consulted a human cardiologist, we could work out a plan similar to one he would devise for a human, leaving only the problem of size. The drug dose for a one-hundred-fifty–pound human would not be the same as that for a one-thousand–pound pig. If a human needed two of these pills, would the pig need eighty? I learned to go to a compounding pharmacy and tell them the poundage of my pig in order to purchase a more concentrated version of the drug. That saved a lot of money, and it worked for my pigs. I may be the only person who has treated a pig with angina by applying Nitro patches to their shaved ears, or who has used injections of Lasix to help remove fluid from the lungs of pigs with chronic congestive heart failure.

Of course, some pig injuries and medical issues pertain only to pigs. Despite their large bulk and short legs, pigs are not clumsy. I was often amazed by how the biggest farmed pigs could quietly and agilely make their way through the woods. It was also a joy to see our younger pigs skip like mountain goats over rocky cliffs. Nevertheless, bruised hoof pads and twisted ankles became an inevitability. In most cases, I would simply apply liniment to reduce any swelling. The injury would have to be severe before I would medicate the animal to lessen the pain. Lessen—but never eliminate—it. Again, this resulted from seeing the world through a pig's eyes. If I were to take away the pain completely, that pig would use their injured limb as if it were healed, turning a simple injury into a major one. Some alternatives are worse than pain, and that applies to humans as well.

A significant percentage of our aging, intact females routinely developed uterine tumors, which were invariably benign but which grew at a prodigious rate. One of our pigs, who had looked merely fat on Monday, by Thursday had such a large tumor that she appeared to be straddling a basketball, with her feet unable to touch the floor. Out of that two-hundred-forty–pound pig, our vet removed a forty-pound tumor. When we had encountered these tumors often enough to consider them typical, we began spaying all our females who were still small and young enough, especially our potbellies.

In my early years as the sanctuary director of Mini-Pigs, most of the calls I had received from other directors had been about practical matters: pig transportation, sanctuary management, fundraising, and so forth. Those I received from animal guardians usually had to do with behavioral problems. Once at The Pig Preserve, these callers more often asked me for diagnostic and treatment advice. The first words out my mouth were always: "I'm not a veterinarian. I don't know your pig, and I don't want to be responsible for giving you the wrong information."

Some of these callers had already been turned down by half a dozen vets, and they couldn't afford to send their pigs to a veterinary college. My quandary became: *Do I try to help these people and risk making a mistake, or do I withhold potentially useful information, perhaps causing their pigs to suffer and die?* That was a difficult decision for me, but invariably, I did try to help. I would tell those two dozen people who called each spring with a list of all-too-familiar symptoms: "I think your pig might have erysipelas. Get your vet to give them penicillin shots for five days, and they should be okay."

Fortunately, in 99 percent of these cases, I later received a call telling me that the pig had recovered.

Both in factory farms and in laboratories, pigs are considered disposable commodities, without rights or feelings or individuality. I can't condone experimenting on pigs to improve the health of humans, though I understand why some people might think it's necessary. I have a heart condition, but if a doctor told me he could save my life by implanting a pig valve, I would not agree to it. Neither would Laura. Slaughtering another animal to extend a human life doesn't seem like a fair or ethical trade-off to us.

Most animal experiments are not necessary. I've watched films in which pigs were deliberately injured to train doctors on how to treat battlefield wounds. While this practice may have seemed justifiable fifty years ago, the advances in today's computer simulations have made the arguments for it invalid. The same goes for using animals to train physicians. No justifiable reason exists for turning pigs (or any other sentient creature) into experimental subjects for this purpose. Computer modeling does as good a job as—if not a better job than—torturing and killing these helpless beings. Student doctors can also learn how to treat victims of catastrophic events by attending busy urban emergency rooms and trauma centers. To their credit, many prestigious educational and training institutions have abandoned these cruel practices in favor of more effective and more humane methods.

Pigs are probably the most denigrated of all farmed animals, as reflected in our language: Ugly as a pig. Dirty as a pig. Smelly as a pig.

Chauvinist pig. Road hog. Pig-headed. Putting lipstick on a pig. You swine! You pig! Sweats like a pig (even though pigs don't sweat). Perhaps, unconsciously, we understand the similarity between ourselves and pigs, and it offends us. It isn't just our shared physiognomy. When you look into a pig's eyes, you can see you are communicating with an emotional, intelligent creature. Since it's easier to kill a being who is "not us," we create an artificial distance between ourselves and pigs so we can breed, raise, slaughter, and experiment on them in horrific numbers.

But they *are* like us.

14

You Don't Eat Your Friends

Before I knew anything about sanctuaries, I was a meat eater . . . a meat *lover.* One day, I was carving a ham with little Patti Murphy wheeling around my feet when, suddenly, a light bulb went off in my head: *What's the difference between this pig I'm carving and the little pig at my feet whom I love so much?* The glaring and obvious answer was: NONE!

Laura came to the same realization at about the same time, so eating pork products became abhorrent to us. We stayed pork-free for about a year before the second realization hit us: *If eating pigs is cruel, what reason can we give to justify eating cows and chickens and all those other sentient beings?* Again, the answer was: NONE!

It took another two years for us to transition from meat eaters to vegetarians to vegans, and that part of our journey was difficult. Back in the nineties, virtually no palatable substitutes for dairy or eggs or meat or other animal products were available. One of the first meat substitutes came in cans, like dog food. When taken out of the can, it looked like dog food. When eaten, it tasted like what I imagined dog food might taste. You could also mix a dried meat substitute with water, then press this into quasi-hamburgers that lacked taste and consistency. Products like tofu were not readily available, and nobody seemed to know how to make them palatable.

Ever hopeful, Laura and I bought each new vegan product as it came out. I was excited when we found some vegan mayonnaise, because I love mayonnaise. I made a big sandwich with juicy tomatoes from our garden, spread mayonnaise on both sides of the bread, then bit into it. The taste reminded me of that time when I, as a kid, experimented with eating library paste. (Okay, in the spirit of total transparency, I ate library paste more than once as a kid and even considered myself somewhat of a connoisseur of the stuff).

What made veganism especially challenging was the fact that I was the cook for fifteen to twenty hungry firemen. Now, many perks came with being the firehouse cook. I was able to choose the meal for each day. I didn't have to wash dishes or do the more odious firehouse duties, such as cleaning the bathroom. I enjoyed being firehouse cook, and I was good at it, but when I tried to get my guys to go vegetarian and later vegan, I faced a rebellion. I had to give up being the cook. Since I didn't want to sit at a table where they might be carving a pig, I would bring my own food to work and eat before or after the others on the shift. This made me seem antisocial in a world where teamwork and being part of the firefighting family were very important.

My firehouse guys were a meat-and-potatoes gang, but when we started to hire female firefighters, some were vegetarian. I think I even converted a couple to veganism. That felt good, because then I could share meals with one or more of my shift mates, removing some of the stigma from being a vegan. When enough palatable vegan products slowly made their way onto the market, the guys occasionally let me make a big pot of vegan chili or a vegetable stew, but this remained a hard-fought battle over at least fifteen years of my career as a fireman.

Visitors to our sanctuary were allowed to bring only vegan products. We publicly advertised: no meat or meat-related products. This policy became easier to enforce as more people learned that a plant-based diet is healthier for them and better for the planet.

Being vegan is more than a diet; it is a lifestyle choice. When we hired Tim Lewallen, one of our requirements was that any prospective worker must be vegan. However, in most areas of rural Southern Appalachia, vegans were few and far between, and our need for extra help was pressing.

As Tim became acclimated to work at the sanctuary, we shared vegan dishes and products with him, most of which he liked. Tim was a magnificent addition to our family, but the simple fact was that Tim couldn't afford to feed his family vegan food. This was a depressed, low-wage part of the country, and Tim was a farmhand with three daughters. Eating vegan in a small Southern town like ours was difficult and expensive. Vegan items were only occasionally available at Walmart, and Jamestown had no vegan restaurants.

For Laura and me to be vegan in Jamestown, we had to travel sixty miles to find a store with vegan products. For many people like Tim who live paycheck to paycheck, eating vegan is prohibitively expensive. For us to have mandated that Tim be vegan on the salary we paid him was not realistic. Because of Tim's love of animals, he told me that he would readily convert to veganism if he could afford it. The fact is he couldn't, and Laura and I couldn't afford to pay him more. Though our commitment to veganism is rock-solid, I don't sit in judgement on people of goodwill who are making the effort.

Laura and I stopped holding weekend open houses when we moved from Virginia to Tennessee. The Pig Preserve was more remote than our sanctuary in Virginia. We had limited parking spaces, and our small town had only one mom-and-pop motel, which was frequently full on weekends. We also had fewer volunteers, though our program for traumatized pigs who needed help to adjust to their new freedom continued.

Poppy, a tiny, silver potbelly pig, came to us terrified of humans. Her previous guardian's husband had abused Poppy over a long period of time, including shooting her with a BB gun for "sport." All our attempts to heal her badly scarred back with its raw, bloody patches failed, and she remained so paralyzed with fear that she became frantic when approached or touched.

The only upside to this part of Poppy's story was that her abuser received a six-year sentence for animal cruelty.

We took Poppy to the University of Tennessee, where, after a variety of tests, we were still unable to pinpoint a reason for her wounds remaining raw or to find a cure. Though our vets weren't sure Poppy could be saved, after a great deal of trial and error, we began treating Poppy's injuries with medicinal honey and several other products designed for burns. Slowly but steadily, fresh, new skin appeared. After several years of constant treatment, Poppy's back has almost completely healed.

Poppy's emotional wounds needed their own special program. To help her to decompress, we kept her in a separate pen, where volunteers read to her every day till she had the courage to interact with humans and other pigs. She has adjusted well to her life in the Baby Pen, where her companions are very young pigs and a few other compromised pigs her size.

A more serious reason for limiting visitors to the preserve, beyond the difficulties of our remote location, was biosecurity. Swine flu, hoof-and-mouth disease, and a host of other diseases made not only me but also our state veterinarian nervous.

Our rescue stories often feature a chain of unlikely events involving people who couldn't imagine they would ever help to save a pig. These stories all seem miraculous to one degree or another, all involving a cosmic intersection of timing, circumstances, opportunity, and unselfish human assistance.

In a semi-rural area of Central Florida, a young Yorkshire pig was being raised for slaughter. Taken from her mother and her litter when she was only a few weeks old, this little pig spent the first year of her life confined to a lonely, tiny, trash-littered pen. Fed on the cheapest, antibiotic-laden commercial pig feed, she soon grew to the ideal "slaughter weight" of three hundred pounds.

On the day she was to be killed, her "owner" hauled her into his yard, then hung her upside down by her rear feet from a large tree, preparing to skin her alive.

A next-door neighbor, drawn by the poor pig's frantic squealing, watched these slaughter preparations with a growing sense of outrage. She was not an animal rights activist. She was just someone who understood that what was about to occur was not right. Grabbing a shotgun, she raced out of her house, then threatened to shoot the pig's torturer unless he released the pig.

The police were called. Undeterred, the woman, who by then was supported by her neighbors, adamantly repeated her intention to shoot the pig's "owner" if he continued to torture and kill the pig as planned. Perhaps surprisingly, the police negotiated on the side of the woman and her neighbors, who took up a collection to buy the pig.

Now that the pig had been saved, what would become of her?

As it happened, one of our sanctuary friends lived near the place where this saga was unfolding. Although she worked primarily with miniature pigs, she offered to give the pig—now named Dee Dee—a temporary home. Because Dee Dee, in her short, horrific life, had never been socialized to humans or to other pigs, she had developed a fear-related habit of biting her human caretakers, making her current situation unmanageable.

Once Laura and I were contacted, we made arrangements for Dee Dee to be transported to The Pig Preserve. Her arrival was similar to that of so many others. As she stepped out of the trailer, she seemed not to believe the sight and smell and feel of all that luscious green grass under her hooves. Nor could she fathom that instead of a tiny, barren pen, she now had one hundred acres of woods, pastures, and ponds at her disposal.

For hours, we watched Dee Dee wander about, hungrily grazing. Eventually, she encountered several of her new herd mates. Her initial reaction was to run from them, but soon she discovered a group of younger pigs her own age. These pigs were being shepherded by an older sow named Tootsie, who had adopted the youngsters on her arrival at the sanctuary. Within a couple of hours, Dee Dee was accepted by her new family.

During the next two years, Dee Dee became a fully grown, healthy pig. She also learned to trust humans. When I went into the pasture to work, she would run to me for a back or belly rub. She could always be found in the company of three other young Yorkshires—Phoebe, Prudence, and Piper. Once called Pigs One, Two, and Three, they had been purchased as piglets from a local farmer by a professor at the University of Tennessee College of Veterinary Medicine. His intention was to do a study on hoof growth for the treatment of lameness in swine and dairy cattle.

For a year, the three sisters lived in a ten-by-ten–foot indoor concrete stall, bathed in perpetual overhead light, in the college's Large Animal Facility, so students could measure and photograph their hooves. When the project was finished, the three sisters were to be sold for slaughter, in accordance with the college's policy.

Fortunately for the pigs, the students had grown fond of these three impish, cute, lovable girls. When the university refused to relax its animal disposal policy, the students hid the pigs in a different stall in the cavernous Large Animal Facility. Inevitably, an officer of the college discovered the pigs, and the students were ordered to take them to auction.

The students did as ordered—sort of. They took the pigs to an auction, but not a farm auction for slaughter. Instead, they transported them to an auction for exotic stock in Cookeville, Tennessee, where they would have more control of the proceedings. When their pigs came up for auction, the students bought them back.

Once again, they hid the sisters in the cavernous vet college, and once again they were discovered. That's when I received an SOS call from Dr. Sarel van Amstel, a world-renowned expert on farmed animals, who was in charge of the hoof project. As it happened, Dr. van Amstel was also one of my go-to vets at the University of Tennessee. When we agreed to take the sisters, he and his wife delivered them to us at The Pig Preserve.

As a rule, large animal researchers are an unemotional lot. They are trained that way. Their institutions demand it. They deal daily with "production" farmed animals, valued solely for their market worth. On this day, I had the rare pleasure of watching this particular researcher grow teary-eyed as the three girls stepped off the trailer, then began jumping and spinning around in the pasture with the pure joy of being freed from a pen for the first time in their lives.

Pigs One, Two, and Three were named Phoebe, Prudence, and Piper, after the three good witches in the television series *Charmed*. They grew into rambunctious, outgoing teenagers with boundless energy and a love of life. Each day, when I fed our potbelly pigs, Phoebe, Prudence, and Piper would come to the pasture fence, where they begged for—and received—three specially made peanut butter and jelly sandwiches.

After being adopted by Tootsie, the three sisters integrated into her growing family of younger pigs that eventually numbered twelve.

Tootsie came to us from the Wiles Hog Farm in Ohio. In 2009, HBO aired a documentary, *Death on a Factory Farm*, using undercover footage shot at this facility that showed piglets tossed into crates from across a room, a sick piglet hit against a wall to be "euthanized," and a sick sow hung by a chain from a forklift until she choked to death.

Tootsie was one of the sows being exploited at Wiles Hog Farm. As a piglet-producing machine, she was destined to bear litter after litter until she was "used up," when she would be sold for slaughter. Thanks to animal activists, she and three other lucky young sows were liberated. After being moved several times, Tootsie ended up at a small animal sanctuary in northern Virginia. Like too many other grassroots efforts, this one failed, which meant Tootsie and the others would be confiscated by the local authorities and either put down or sold for slaughter.

After hearing of Tootsie's plight, The Pig Preserve offered her, along with several other pigs from this facility, a forever home, then paid for her transport to Tennessee.

As a sow valued solely for breeding, Tootsie had never had the luxury of actually raising her piglets. For a few short weeks of nursing, she had been forced to lie on her side crammed in a gestation cage, feeding her piglets through cage bars. After each brood was taken from her to be raised for slaughter, she would be artificially inseminated to produce another litter.

Tootsie became The Pig Preserve's surrogate mom, adopting all of our young pigs. Still haunted by all the abuse she had suffered, she preferred to keep her distance from humans. Over time, Tootsie did accept me, because she associated me with her daily food. She even let me interact with her adopted children. Otherwise, she remained a bit aloof and reticent with strangers.

Not too long after Tootsie's rescue, we were contacted by some animal lovers with an urgent request to take in Momma Fleurri, an old sow marked for slaughter by a New England museum now that she was no longer useful for breeding. As with Tootsie, brood after brood of piglets had been taken from Fleurri soon after birth. In an upbeat chapter of this familiar story, members of Fleurri's community, who considered her sort of a "pet," asked if she could be spared. Before the museum would agree, a permanent home had to be found for her.

A community search spearheaded by a local radio celebrity, who happened to be vegan, led to our sanctuary. We decided to let Fleurri, who was elderly and overweight, come here to live out the last few years of her life in peace and tranquility.

In early spring, Fleurri made the long trip from New England to Tennessee in a large stock trailer. In spite of having seen her in photos, I was not prepared for the gigantic white pig who lumbered off the trailer into our barnyard. She was a massive Yorkshire, with relatively short legs and a huge body. And while not aggressive, she was definitely not a "people pig." Without any fanfare, she made herself at home in our main barn. Not a single resident pig had the courage to challenge her.

Fleurri was an enigma to me. Though not in the least aggressive, she seemed as disinterested in the companionship of other pigs as that of humans. I began to wonder if we had made a mistake in accepting her into the sanctuary, since she certainly did not seem happy.

A little over a month after Fleurri's arrival, we took in three very young female farmed pigs—two Hampshires and one Yorkshire, named Ciara, Rosie, and Rosebud. Though all three were from separate rescues, they arrived together on the same spring day because we had arranged for them to be fostered over the winter.

While they were still in their trailer, we went through the usual unloading process, which meant checking them for mange mites or hog lice and ascertaining their overall physical and mental condition. After that, the rear gate to the trailer was dropped, allowing the new pigs to find their way out into the barnyard at their own pace. Typically, this event drew a fairly good crowd of porcine onlookers, resembling rubberneckers at the scene of an auto accident or house fire. And of course, the sanctuary dogs had to be involved in the "meet and greet," adding a cacophony of barks to the mildly chaotic barnyard scene. We humans also stood by to referee, in case the greeting committee became overly exuberant, and to document the new arrivals with a digital camera or phone.

All was going smoothly, until a couple of the new girls let out a series of amazed and startled high-pitched squeals. Suddenly, from around the back corner of the barn came a series of deep, guttural grunts, followed by the massive form of Fleurri, rounding the corner like a charging rhinoceros. Head down, eyes ablaze with fury, and snapping jaws full of slobber, she raced towards us at an impressive speed and with great agility, considering her huge bulk. Everyone—dogs, cats, pigs, and humans—scattered for safety, leaving the three girls standing alone in the middle of the barnyard, wondering what was happening.

When Fleurri reached the three newcomers, she slammed on her brakes, then whirled around to challenge the assembled group, not that any of us had any thoughts of interfering with whatever was about to happen next. She then turned to the three new youngsters, gave them a good going-over with her snout and eyes, uttered a deep "oof," then turned and walked into the barn with the three youngsters trailing closely behind her.

From that moment on, Ciara, Rosie, and Rosebud became Momma Fleurri's adopted children. Everyone—human and pig alike—quickly learned not to do anything to make any of Fleurri's chosen three squeal in anger or fear. This would immediately bring an enraged Momma Fleurri charging from wherever she had been loitering to defend *her* babies.

They were never out of her sight. On warm spring and summer days, Fleurri could be found sunning herself in the deep grass on the side of the pasture, while Ciara, Rosie, and Rosebud climbed all over her, then snuggled under her at nap time. The youngsters enjoyed having a mother, and Fleurri was finally able to raise a litter of young pigs in place of all those taken from her.

Momma Fleurri also underwent a complete change in attitude and demeanor in regard to the sanctuary. She became more outgoing and social. Maybe, for the first time in her long life, Fleurri was happy.

I wonder how many people who have visited a petting zoo have taken the time to consider what happens to the piglets and other baby farmed animals once they outgrow the "cute" stage. The stark truth is this: As soon as they are too large to appeal to young children, they are summarily shipped off to be slaughtered.

Bimbo and Betty Boop were two such pigs. Located at a petting zoo in North Carolina, they were fast approaching the age at which they would be "replaced" by younger piglets. A college couple who had visited the pair at the zoo made it their mission to save them from slaughter. Without a long-term plan, they purchased these pigs, then transported them to their urban home, where they immediately came under intense pressure to remove the "livestock" from their premises.

In desperation, they contacted The Pig Preserve.

Though strapped for funds, we agreed to take the two youngsters if the couple could transport them to us from North Carolina. They rented a van, then drove Bimbo and Betty Boop over the Smokies up into the Cumberland Mountains. I am sure that both pigs found The Pig Preserve a major improvement over their previous tiny pen and the constant attention of strangers poking at them.

Bimbo grew into a handsome Oxford Tan and Spot pig, with large black splotches on his tan hide and an attitude as big as all outdoors. Betty Boop developed into a svelte and sweet Yorkshire. Both enjoyed romping with a herd of youngsters. Bimbo developed the devilish habit of trying to dash between my legs at feeding time in order to arrive a nanosecond earlier at a trough. This was amusing when he was small. When he no longer fit between my legs, he would barge right through anyway, lifting my feet off the ground and sending me rolling off his back into a rush of other pigs eager to get to their feed bowls.

Bimbo considered this great sport.

Colleen, a beautiful, young white female, was abandoned on a busy highway with three other young pigs, one of whom was a boar. The pigs were rescued by a kindly family who agreed to hold them until we could transport them to the sanctuary. When it became apparent that Colleen was very pregnant, she was rushed here by truck. While three miles from the sanctuary, she delivered five adorable white piglets. We welcomed mom, dad (now neutered), and the piglets to the sanctuary.

Clancy, a young boar, was discovered in a three-by-three–foot pen, living in his own filth and existing on rotten fruit and moldy horse feed at a "wildlife sanctuary." Someone had tried to skin him alive, leaving him horribly scarred and missing an ear. At the sanctuary, Clancy healed physically but remained fearful of humans. While he would accept hand-held food, he would not let anyone touch him. Fat and sassy with other pigs, Clancy took to hanging out with several older males in a "guys' club" in their special part of the pasture.

Aberdeen was rescued from a factory farm at six months old, after having already been sorted by weight and spray-painted to indicate that he was ready for market. Like other piglets born into the factory-farm system, he had had his needle teeth, testicles, and tail removed without anesthesia at only weeks old. Human attention to him meant anything from disregard to outright cruelty.

Penelope, a forty-pound, partially crippled female, came to us pregnant and close to starvation. Because of her condition and small size, our vet determined that delivering her babies would probably result in Penelope's death as well as that of her piglets. After undergoing abortion and being slowly nursed back to health, this loving little girl with gorgeous blue eyes joined a herd of small pigs, most of whom had disabilities.

MacGyver was captured after he had run wild for three months in a suburban neighborhood. The "owner" of this little boar had beaten him repeatedly with a two-by-four to drive him off. When he refused to leave, the "owner" shot him in the stomach, leaving him with a festering belly wound and a terror of humans. Using an intact female as bait, rescuers enticed MacGyver into a pen. Once at our sanctuary, he was neutered and had his tusks trimmed and his belly wound repaired. After months of rehabilitation, he joined a herd of fourteen other pigs. Like Clancy, he healed physically while remaining distrustful of humans.

We named him MacGyver because of his ability to live by his wits for three months, resisting capture.

Rescuing an animal was only the beginning of every story. Once the trailer brought a pig through our gates, we became responsible for their safety and care for the rest of their life, which might be another fifteen to twenty years. I tried to demonstrate to each pig that not all humans were like those they had experienced before coming to The Pig Preserve. I showed each pig that they were safe to trust humans. It became my goal to heal each pig's physical injuries as well as to invest countless hours in healing their emotional wounds. I showed them every day that they were safe, loved, and respected as fellow sentient beings. We encouraged them to wander in the woods, to graze in the fields, to root and forage, to swim in our ponds, and most importantly, to live their lives in the company of other pigs.

Thanks to the awareness campaigns of activist groups like Animal Save, many more people are beginning to connect the meat on their plates with the cruelty of factory farming. We see this awareness reflected in the greater willingness of authorities to lay criminal charges for animal abuse, despite the pushback from those who profit from animal exploitation. We see it in the willingness of some global courts to recognize other-than-human animals as sentient beings deserving of compassion and protection, not as "livestock" to be processed.

I have to believe that one day—perhaps long after I am gone—there will be no more human meat-eaters and no need for sanctuaries like The Pig Preserve.

15

A Death in the Family

Over the years, running The Pig Preserve became a lot like running a giant hospice. I knew that every animal who came through our gate was going to die here. My sense of finality began with the opening of the trailer and a pig's first walk of freedom, because I understood that we were making a birth-to-death commitment of unknown duration to that pig. We usually didn't know the genetics of the pigs who came to us or the details of their lives. Some looked like they'd die fairly early, and they did. Others surprised us by living into their late teens or twenties. As mere caretakers, we had absolutely no say in that. Likewise, some humans don't make it into their fifties while others reach a century and beyond, as a glance at the obituaries in any local newspaper would show.

That life-and-death mystery helped me become a better caretaker when I was feeding and watering our pigs twelve to sixteen hours a day, seven days a week. I did it when I had the flu. I did it when the weather was crappy. I did it on days when I was frayed and required every ounce of my willpower to drag my sorry butt out to the pastures with buckets to get the job done. On those days, it was human nature to be short-tempered, but when I felt snappish, I thought about the pigs I had lost, especially those I had lost unexpectedly whom I had perhaps found in their favorite place after some catastrophic heart attack. That made me stop to realize that I had to treat each pig in my care as if it were that pig's last day on Earth, because too many times, it was.

I felt an even greater sadness when I had to put down a pig, especially one I had raised from a young age. I did it for thirty-six years. I kept thinking it ought to get easier, but it never did. I always asked the vet to let me give the shot that would sedate my pig for the euthanasia. I felt that was my responsibility as the pig's caretaker. Then, afterwards, I would bury that pig with my own two hands, kneeling to say my goodbyes in private over that pig's grave. I would thank God for the privilege of knowing, loving, and caring for that pig. I would also say a prayer for the millions upon millions of pigs and other innocent animals whom I could not save from untimely slaughter.

Despite my own grief, it took me many years to understand that pigs also needed to grieve the loss of their loved ones. I had watched how Ellie Mae had buried her four dead babies—two stillborn and two who died shortly after birth—how tenderly she had carried each of them in her mouth, how she had dug their graves with her snout then lain on their graves, making soft grunting sounds.

Through the years, I also saw that no pig died alone except by choice.

Siblings Wilbur and Henrietta remained closely bonded during their thirteen years with us, after being rescued as piglets from a "farm-to-table" exhibit at a historical museum. During sweet Henrietta's final week, Wilbur lay beside her for two solid days, refusing food. Members of Henrietta's social group also joined them. On the night of Henrietta's death, four of her pig friends formed a square around her. After they left, the larger social group, with whom Henrietta used to feed, dropped in one by one to visit, as if in a funeral procession.

A month later, Wilbur also died, probably of a broken heart.

Fergie was a very benevolent, very dominant, very large farmed pig with a social group of some twenty pigs. When she became terminally ill, she made a nest on the side of a hill in a shady place near a pond, where she would spend her days. Sometimes she'd join her herd to eat, and sometimes I'd take food up to her. As her appetite declined, I tried to find foods she'd like. Maybe I'd make a pan of cornbread for her or mix applesauce in with her feed.

Every time I took food to Fergie's nest, two or three members of her social group would be standing guard over her. And the group rotated, having some type of roster so that Fergie was never by herself during the three weeks in which her health declined. Her guards would be either sleeping beside her or standing by, as if holding a deathbed vigil.

During Fergie's last day, she returned to her barn to die. The adult members of her group, acting as sentries, barred me from entering. They weren't aggressive. They simply created a wall that I couldn't breach, though I was still able to observe.

Starting with Fergie's second-in-command, the pigs would go into the barn one by one, from the most dominant to the least dominant. They would stand solemnly, making soft grunting noises and that distinctive mouth-breathing "whooping" sound. Some would gently prod Fergie's body with their snouts.

The younger pigs appeared confused and rather distraught by this process. They would go around the body, prodding Fergie as if coaxing her to get up.

When all members of the group had finished their goodbyes, they left the body. Five or six of the pigs who were closest to Fergie then disappeared into the woods.

That, for me, was when grim reality set in. With a human death, the funeral director whisks the body away, and the next time you see it, it's encased in a shiny casket, carefully made up, and wearing a good suit or dress, ready for burial. Respect for the dead is ingrained in me, but there's no dignified or genteel way to get a 1,200-pound pig out of a barn and up a hill to dispose of their body. I had to take the tractor, and ropes and chains, and drag Fergie out of the barn, then scoop her up with the bucket of the tractor, drive her to a gigantic pre-dug hole, and, as gently as possible, roll her into it. Typically, and especially in summer, I would pour a couple of bags of lime over the body, push the earth on top, then compact it by running the tractor back and forth so coyotes wouldn't try digging it up.

That is not the kind of funeral we envision as humans, but it was all I could offer, along with my prayers. That and fresh hay in the barn, after pitchforking out the bed where Fergie had died. In most cases, a dead pig's mourners wouldn't come out of the woods until the body had been removed. When I went into the woods looking for them for a wellness check, I'd find them lying down together, not much interested in food. In a few days or maybe a week, they would come back to their barn, and life in that particular group would slowly return to normal.

Fleurri, our old sow from the museum in New England, stayed with us for a couple of years until she was diagnosed with a rapidly growing uterine tumor. Given her age and her massive size, we knew she would not have survived abdominal surgery, so we took the difficult decision to euthanize her. Afterwards, we allowed her body to remain in the barn overnight, so her adopted daughters and the other members of her group could grieve for her. When I checked on them later that evening, I saw Rosie, Ciara, and Rosebud lying on and beside Fleurri, crying huge tears of grief, while the other pigs sat or stood quietly at a respectful distance, keeping a vigil over her enormous body. Early the next morning, I took Fleurri out of the barn and buried her in our cemetery in the woods behind the back pasture.

The pig grieving process was something I learned to respect. Not all social groups behaved the same way, and the form of grieving seemed to be dictated by the dying pig.

Throughout human history, some queens have ruled with an iron fist while others through the strength of their personalities, and I've also seen that among our pigs. Some dominant females retained their position as top hog by booting up each of their underlings every day. Others could command loyalty with just a grunt or a nonverbal gesture, and those queens had the happier social groups.

Patti Murphy was the queen among queens. She possessed an aura that made other pigs naturally want to defer to her. Early in her life, she became used to the fact that a constant flow of pigs would be invading her world. As Laura and I watched with bemusement, she would go over to each newcomer and introduce herself. There would follow a five-minute exchange of verbal and nonverbal greetings, after which it became clear who was in charge.

Patti's final social group probably contained about thirty pigs. She could walk to a trough where twelve of them were eating and, magically, a space would open for her. I never saw her have to push for anything to which she, as a mature pig, felt entitled.

In 2006, as Patti came to the end of her lifespan, she seemed to want to spend less and less time with her social group. I made a big nest for her in my tool shop, reached by a ramp, where she could enjoy her solitude. She would come into the house for maybe a couple of hours, then join her friends for dinner, before returning to her nest to sleep by herself. Meanwhile, Shamrock, her second-in-command, became the alpha pig in what was a totally natural transition.

Like their human counterparts, female pigs have a heat cycle about every twenty-eight days, and they often have menstrual cramps, putting them off their feed for a day or so. They also go through menopause, with their heat cycles becoming more infrequent and unstable, leaving them vulnerable to certain medical problems. Patti began having uterine infections more often. When she lost her appetite, it was her muscle mass that melted away. During what would be her last months of life, she isolated herself entirely from her social group. I believe that on some subliminal level, she knew she was dying, and this was how she wanted to go.

Our vet did blood work and an ultrasound on Patti. We were concerned that she might have cancer. Since a diagnosis was difficult, we agreed the only way to find answers was through exploratory surgery. Patti was still healthy enough for me to be fairly comfortable with her chances of surviving the operation. If cancer was discovered, I knew it would be best to let Patti go while she was still under anesthesia. To bring her back would be to sentence her to a miserable end of life.

I have often been guilty of keeping alive a terminally ill animal for longer than was reasonable because I was imposing human standards of death on them. Instead of releasing that animal from suffering, I wanted to stave off the pain of my own loss by selfishly telling myself, "It's not their time," which was only human. I knew that I had to accept the possibility of Patti's death from her point of view—through the eyes of a pig. Pigs aren't afraid of death because they don't understand it, and they can't fear something of which they have no concept. Pigs live in the present. When they're suffering or in pain, they understand that pain as never-ending.

Patti's exploratory surgery found cancer riddling her bowel, liver, and just about everything else, so we let her go. That decision was a heart-wrenching one. Patti Murphy: Laura's and my first pig—in many ways, a founder of The Pig Preserve.

I always brought pigs who died during surgery back home for burial, unless their bodies had gone for necropsy. Since veterinary necropsy labs are not sterile, I couldn't risk bringing back a communicable disease to The Pig Preserve. In Patti's case, we compromised with a cremation.

Patti's loss wasn't felt too deeply by her social group because she had self-isolated for a number of months, and the other pigs had grown used to Shamrock's leadership. As I look back, it almost seemed as if Patti were weaning them from her because of some porcine premonition or innate wisdom of what was appropriate.

Laura and I were devastated.

No matter how inured we had tried to become to sanctuary deaths, some still clawed through our defenses. Patti's death took us back to that time, eighteen years ago, when we had first brought her home and discovered that she was not a funny-looking dog. And then, to that time, a year later, when she had had her accident jumping off our bed, and we had thought we might lose her. Now we *had* lost her, and it felt like losing a child, one whom we as parents had loved and nurtured for so many years.

ENDINGS

16

A Snapshot in Time: March 2021

Roll Call: 162 Pigs

As the years passed by at The Pig Preserve, our porcine residents settled into five distinct neighborhoods. Some were attracted by the geography—a favorite pasture, a tempting mud hole, a shady forest. Some came with friends and stayed with friends. Some sought solitude. Some sought engagement. Some bore deep emotional scars and needed support. Some were elderly or disabled. Some were babies taking their first exploratory steps. All gravitated to the neighborhood that suited them best.

Homestead Hangout

This community, encompassing the preserve's largest barn, is the most diverse: all shapes, all sizes, all ages, all breeds, and all colors. It has rolling pastures, shady trees, and—a local architectural wonder—a giant mud hole that is still being excavated. Its residents, many of whom you've already met, are friendly and outgoing.

There's Abbey Christopher, the shaggy black Berkshire who narrowly escaped being someone's Christmas ham. The three Berkshire brothers—Wynken, Blynken, and Nod—who arrived emaciated and covered in parasites from their life in the wild. There's Timbo—white, with heavy drooping ears and the long snout typical of the Landrace breed. He spent the first part of his life as a breeder boar, but one so tender-hearted that he grieved every time the piglets in his litter "disappeared." Then there are the Hampshire Yorkshires, Sissy and Bubby, who were too sweet-natured for their homestead family to eat as they had planned.

We call Rosebud a "timeshare" pig because she splits her time between two different barns. She's the only surviving member of The Cougars—the three matrons who liked to collect younger males around them. We lost Ciara during a hospital stay to remove a tumor, and Rosie two summers ago, but Rosebud still has it. No matter where she goes, her loyal boyfriends—Pineapple, Abbey Christopher, and Yankee Wilbur—follow her.

Irish is the little Red Wattle Hampshire mix saved from a flea market where he languished in a wire dog cage marked MEAT PIG. Easter was once advertised as the $80 centerpiece of a festive barbecue dinner. Her best pal is Sally Mae, a Gloucestershire Old Spot, predominantly white with black spots. Sally Mae was to be butchered after she had grown too big for her star turn at a county fair. Rhett and Scarlet, brother and sister Red Wattles, spent three years in a filthy, dilapidated twelve-by-twelve–foot pen steeped in waste-saturated mud till vegan rescuers stepped in.

Remember our outlaws, Butch, Sundance, Toot, and Puddles? They're the four Captain Cook Miniatures once known as The Demons because of their exuberant piglet antics.

Damien and Sunshine, the two ferals who wrecked the trailer transporting them from Florida because of their terror of confinement, now have plenty of room to roam. So do The Little Rascals: Alfalfa, Stymie, Buckwheat, and Spanky, four feral–potbelly siblings also from Florida. And let's not forget Rhonda and Ruby, who arrived at the preserve with their brother Rufus—three mature farmed pigs so pitifully emaciated that you could see every bone in their bodies. Though now ancient by pig standards, Rhonda and Ruby are still with us and still doing well.

Piglet Pasture

This small pasture serves as a safe haven for babies not yet ready to join one of the larger herds; for special-needs pigs; for those requiring extra socialization; and for those recovering from trauma, illness, or injury. Then there are the permanent residents who just happen to like this beautiful spot with its mud hole, its grass for grazing, and its many shade trees, including a couple of apple trees. Since Piglet Pasture is near the house, it's easy for us human caretakers to keep watch over those needing special care.

Luna Rose is the tiny Yorkshire piglet who was thrown onto the dead pile at a factory farm because she was considered too small to be worth feeding. Journey was once the newborn feral who watched her mom and siblings being slaughtered by hunters and their dogs. These two are growing up together, and that's probably how they'll graduate from Piglet Pasture.

Little Tucker is a Gloucester Old Spot, mostly white with black spots. He was taken as an infant from his mother to be hand-raised for slaughter by a child who belonged to an agriculture club. Instead, Tucker's family asked the preserve to accept their little guy, who suffers from dwarfism. He doesn't see well and has a slow gait, yet he fits in perfectly in Piglet Pasture, where he's the biggest kid.

Hamlet, a black potbelly, arrived here as a victim of domestic abuse who was very fearful of people. He also suffered from overgrown hooves, with one actually twisted into a corkscrew. Now that he's beginning to feel safe, he's letting down his guard to let in some love. Jojo, another black potbelly, lost his happy home in a hurricane, and his family hasn't yet been able to take him back. Meanwhile, we see that he gets his fair share (and more) of belly rubs.

Poppy, you may remember, was used as target practice by an "owner" with a BB gun, who is now serving six years for animal cruelty. She's recovering in her own time and learning a lot about literature from the volunteers who read to her. Jethro, a black potbelly, had what seemed like a peaceful, happy home until it suddenly became violent, with guns and fury. Though he's still afraid of touch, he also responds pleasurably to the human voice.

Peaceful Pond

The pigs who choose to live here are usually the curious, social kind, who like to amuse themselves by watching humans work. Since they're accepting of newcomers, this is the place where graduates of Piglet Pasture, like Luna Rose and Journey, often settle. This large neighborhood has plenty of pasture and some woods, and easy access to the upper pond with its spring-fed mud hole. Since its residents are close to the house, they're the first to greet visitors, and—a special perk!—they're also the first to see when feeding begins.

Remember Aberdeen? He was the six-month-old white Yorkshire rescued from a factory farm where he was already spray-painted to show that he was the right weight for slaughter. Puppy, a female Hampshire, was yet another rescue from one of those high-school Future Farmers of America programs. Fortunately, Puppy's student chose to save her beloved hand-raised friend rather than enter her in a show for a blue ribbon, which would have ended with Puppy being sent to slaughter.

Oreo, a large black potbelly–feral–farmed mix with a chaotic background, shares a barn with Lulu, another potbelly–feral mix. They're a peaceful pair, welcoming to all visitors as laid-back as they are.

Village Vista

Many of the residents of Village Vista, which is tucked in the woods, arrived in closely bonded groups. The Duroc Yorkshire mother of our three sisters—Buttercup, Iris, and Flower—had been featured in a zoo exhibit. When she became pregnant as a result of her handlers' carelessness, the zoo owners decided to send her to slaughter. She was rescued by the zoo's veterinarian, who took her home, then presided over the birth of her large litter. Momma, along with several of her brood, is still with the compassionate veterinarian. The rest, like Buttercup, Iris, and Flower, were placed in sanctuaries.

You've met Silico, the baby feral who was found at roadside covered in ticks, then was bottle-fed by his rescuers till he was big enough to be transferred to the preserve. His best buddies in Village Vista are farmed pigs Arky, Biaggio, Charlie, and Chester. Together, they have bonded to form their own guys' club.

Chicory, Aster, Clover, Periwinkle, Dandelion, and Violet—named after plants native to Tennessee—were the cull from a factory-farm breeding program and scheduled for slaughter till an employee intervened. They're Meishans, an Asian breed, much smaller than European and American farmed breeds. They're also more docile and will lie in your lap, eager for belly rubs or ear scratches.

Lulabelle, a mottled black and white feral, was accidentally caught in a humane trap that was intended for the relocation of raccoons. The trap's kindly owner cared for Lulabelle for many years till he fell ill. She arrived at the preserve grieving her lost human and intimidated by the other pigs, but she is now making a comfortable adjustment.

Remember Big Bad Bobby, who once used his massive size to intimidate his caretakers and any other pig who crossed his path? When Bobby discovered how good life could be at the sanctuary, he stopped acting out. He's a Village Vista resident, along with his pal—shy, lovable Blue Eyed Wilbur, who grew up in isolation, believing he was the only pig in the world.

We've also told you about Ella, a black Hampshire with white front legs and shoulders. She arrived at the preserve with a big chip on her shoulder and was snappish and ill-mannered towards humans and other pigs. The Pig Preserve's reading program helped soften her edges so that now Ella, too, has found friends.

Hidden Hollow

This neighborhood includes the preserve's most densely wooded area, along with some pasture, a large mud hole, a creek, and easy access to ponds. It's the favorite of many of our ferals and some of our large farmed animals, who often make day nests in the forest and are avid foragers.

Here live our four ferals who were named for activists: Memphis, Martin Luther, Coretta Scott, and King. So does Momma Memphis's surprise brood, also named for activists: Buddha, Corrigan, Dalai, Gandhi, Jane Goodall, Nelson, and Anita.

We have another feral family group: siblings Charity and Chipper, along with Charity's offspring Joey. Then there are Jack and Jill, rescued from a failed sanctuary in South Carolina. All are super-friendly feral ambassadors. Charity is the matriarch who stands guard over the rest; Joey is a crowd favorite.

Among Hidden Hollow's farmed animals, we also have Momma Miracle and Baby Miracle, who are Gloucestershire Old Spot–Yorkshires. Momma is spotted black and white, while Baby is pink. Their best friend is Millie, a pretty, red Hampshire–Red Wattle. All three get along well with Charity and her relatives, creating one big happy Hidden Hollow community.

17

A Time to Say Goodbye

It began five years ago.

That was when problems with my heart resulted in my going into the hospital to receive a pacemaker. By then, I was also having trouble with my knees and my back: more warning signs that I was dealing with an aging body. Both my physician and my cardiologist were urging me to retire.

Laura and I knew there was a timeline. We just didn't know when the end of that timeline might come. Finding dedicated inheritors to provide love and safety for our pigs was not going to be easy. As the months and years passed, bringing with them a few false starts, that timeline shortened while the intensity built.

Laura had realized, sooner than I, the toll that running the sanctuary was taking on both of us. With greater foresight, she had jumped off the bandwagon before me, creating a philosophical wedge between us. As the thin thread holding up the Sword of Damocles over our heads weakened, the reality of our situation became more daunting: If something catastrophic happened to me, Laura and Tim would be left to look after the pigs in a fire-sale situation. I had already seen that scenario played out, tragically, with too many other grassroots sanctuaries, but what were we to do?

I had spoken to Wendy Smith a couple of times on the phone about minor issues in regard to the health of her pigs. Wendy and her husband Josh owned a five-acre sanctuary called Odd Man Inn, located in Seattle, Washington. Though they were open to rescuing all farmed animals, they specialized in pigs. After five years, Odd Man Inn had about thirty pigs, two llamas, two cows, and a couple of goats and sheep. Amazingly, the Smiths had also adopted out some four hundred animals to good homes!

My friendship with Wendy deepened during our back-and-forth conversations about a serious health issue with their pig Jolene. This bonding resulted in Wendy's paying a visit to The Pig Preserve last October. By her own account, she was blown away by the beauty of our place and felt a sense of wonder as she watched our pigs peacefully foraging through the woods, grazing across the pastures, wallowing in mud holes, and napping under shade trees while also serving as goodwill ambassadors in their eagerness for treats and belly rubs.

Over a couple of months, a deal was struck for the merging of the Seattle sanctuary with The Pig Preserve. This would involve Odd Man Inn purchasing our Tennessee property and adopting all of our pigs. In April, with the contracts finalized, we signified the turnover by posting a new sign on the Tennessee property: ODD MAN INN.

As I write this, Wendy is closing up the Seattle sanctuary and preparing their animals for the big transition. Josh's trailer and camper are parked in our front yard, allowing him to put in long hours as he's being pulled in forty-seven different directions. Possibly, he's feeling a bit overwhelmed, coming from five acres to one hundred acres with an additional 160 animals, just the way Laura and I felt when we started The Pig Preserve. That would be a natural reaction, but Josh is a highly motivated, well-focused, Type A kind of guy. He'll also have Tim, who knows the lay of the land and the ins and outs of all the machinery, to help with the transition.

Laura quit her hospital job to box up our personal possessions, for which she has turned into a Mr. Clean sort of person. When we founded The Pig Preserve, I promised her retirement while we were still healthy enough to enjoy it. She is already dreaming of a modest waterfront property somewhere in the Florida Keys, with a boat for excursions and not for hauling pigs. In a couple of days, friends and relatives will be descending on the house to complete the pack-up so we can vacate it for the Smiths. For the summer and fall, we will be moving temporarily to Tappahannock, VA, where my daughter Jennifer and her husband are graciously allowing us the use of their large RV and lot on the Potomac River while we search for that Keys rental.

I started my emotional and physical withdrawal from the pigs by putting distance between them and myself. I am not nearly as involved with them this week as I was a week ago. That feels strange. Not yet in a good way. The pigs are fine. I'm the one suffering separation anxiety.

When I was a Marine Corps company commander in charge of maybe 250 men, about 10 percent were problem guys and 10 percent were stellar performers. It was that 20 percent whom I came to know intimately, and that's the same with my pigs. I have my favorites—ornery ones who turned into smooches, sweet ones I raised from piglets, special-needs pigs requiring more hands-on care. Those are the ones whom it's harder to deal with on a daily basis when I know I'll soon be saying goodbye.

All that is personal.

On the philosophical level, I'm overjoyed, relieved, and reassured to know that a new generation will be taking over Laura's and my legacy, with the resources to see that our pigs will not be jeopardized by my age-related maladies.

Immediate plans for the sanctuary call for a complete remodeling of the house. The leaky roof, the repair of which was never high enough up on Laura's and my priority list, will now be replaced. Bunkhouse-style lodgings will be created for volunteers and visitors. The animal barns will be remodeled. Fenced areas will be created for the cows and goats and sheep, until those animals can be safely integrated into the sanctuary.

In a couple of months, Josh will return to Seattle to make several 2,700-mile trips to bring supplies and equipment to Tennessee. After that—probably in July—he'll start bringing the animals. When the last one has left Washington, Wendy will follow.

After this transition has taken place, I will have no formal connection with the sanctuary. By a gentlemen's agreement, I will make myself available as a sounding board in regard to the medical history of various pigs, Tennessee's zoning regulations, and anything else that might pop up. In five years, I imagine Odd Man Inn will be a premium sanctuary—bigger, more inclusive, and more modern than The Pig Preserve. As I've said before, a sanctuary has to reflect the personalities and values of its owner-directors. I expect Wendy and Josh to make this Tennessee sanctuary their own. I don't expect it to become "Richard's sanctuary on steroids."

I am not yet looking forward to life in the Florida Keys with the same passion as is Laura. I am going to need something more demanding to do than boating, reading, and drinking margaritas while watching the sun go down. All the Keys' fire stations are run by career firefighters. I know from past experience that they're not going to want some old geezer dropping in to say, "Hey, I used to be a firefighter in Virginia back in the nineties," in hopes of a charitable chat over a cup of coffee.

I know of a couple of small sanctuaries in the Keys—one for birds and one for turtles. The turtle sanctuary actually cares for the whole gamut of marine animals, not just turtles. Perhaps Laura and I could volunteer there.

I would like to retain some contact with the network of pig sanctuaries, where I still have so many friends. The rescue of a pig from an abusive situation often falls through because no one has the time or the resources to deliver them to their safe place. Commercial transport tends to be very expensive for grassroots sanctuaries, and too many shippers are callous about the animals. If I could come up with a truck-and-trailer combination, perhaps I could make myself available to do transports for sanctuaries from, say, the East Coast to the Mississippi, for just my expenses.

It's bittersweet to be leaving something that I've done for thirty-six years and that I've loved. The "bitter" comes from the loss of my intimate relationships with our pigs, along with the accompanying sense of purpose. The "sweet" comes from knowing that the promise I made to our animals—that they will be well cared for and well loved—will be kept, and that the sanctuary that Laura and I poured ourselves into will evolve under a new name and different ownership. I'm even more comforted by the knowledge that pigs everywhere can count on the continuous effort of activists advocating against factory farms.

While working at The Pig Preserve, we considered it our primary mandate to educate the general public on the suffering of animals in the factory-farm industry and on the health benefits of a plant-based diet. As our reputation as a high-quality, innovative sanctuary grew, we also became a magnet for animal activists. Since virtually all of these activists are vegan, I at first felt that hosting these groups was a waste of time, like preaching to the choir. It wasn't until I met Anita Krajnc, the founder of the Animal Save Movement, that I learned how vital visiting The Pig Preserve can be for animal activists.

In December 2010, Anita, who lives in Toronto, Canada, created Pig Save after seeing the suffering in the eyes of the pigs in transport trucks that passed her on the way to a nearby slaughterhouse. Later, Anita added Cow Save and Chicken Save to form what is now collectively known as the Animal Save Movement, a global phenomenon with over nine hundred chapters the world over.

Members of Animal Save hold vigils in which they bear witness to the helplessness of animals inside the trucks taking them to their inevitable deaths at the slaughterhouse. Where possible, activists give water to these animals, who are often either dehydrated in the scorching heat or on the verge of freezing to death. They hold up signs to pedestrians and motorists, protesting the suffering of these innocent creatures. They post photos and videos on social media, which are sometimes also picked up by TV stations and newspapers. In this way, they force public awareness of the savagery of the factory-farm industry.

On Anita's first visit to The Pig Preserve, she explained to me: "At our vigils, we see the pigs during their final hours of life. Most are experiencing sunlight for the first time through the breathing holes in the trucks, perhaps raising hope in them for something better. They're cramped together and often covered in feces. Some are sick and dying. Some have been injured. We can't prevent the horrors awaiting them, but we can allow them a touch of human kindness. Exposure to these atrocities is very draining, and sooner or later most of us suffer burnout. When we come to a place like The Pig Preserve, we recharge our batteries."

Because I had never attended a vigil for pigs on their way to slaughter—never wanted even to think about it!—I needed someone to explain that to me. After my conversation with Anita, whenever we received a call from activists telling us they had just been to a vigil in the area and asking to visit us, I would say, "Yes, we'll make the place available to you."

At the preserve, these activists would lose themselves among the pigs, experiencing the peaceful serenity of this place. They would see the pigs happily going about the business of just being pigs. This helped to offset the negativity they had experienced and to provide a purposeful context for their actions.

I couldn't do what these activists do—stand beside those transport trucks with determination and forbearance. With my Irish temper, I'd be dragging drivers out of their seats, and I'd be in jail within fifteen minutes. Most activists likely couldn't do what Laura and I have done for thirty-six years either. All of us "bear witness" according to our own talents and opportunities. We form a natural partnership in our struggle to eliminate the unconscionable factory-farm system and to save the animals who can be saved.

It is my belief that those precious few are spared to become ambassadors for their species. When our gentle giants interact with sanctuary visitors, it is almost as if they know they are special, creating bonds of love. That is why I believe their stories must be told for people who will never have the privilege of entering their world—who will never experience the depth of pig intelligence, or laugh at their inquisitive nature, or admire their social skills, or enjoy their unique personalities, or share their hedonistic love of a cool mud bath on a hot day and their messy delight in eviscerating a ripe watermelon.

So, if I ask myself, looking back over the last thirty-six years: *Would you do it all again?*

Yes. Unequivocally. In a heartbeat.

IN MEMORIAM

Patti Murphy. Big Earl. Damien "Stinky" Gilhoolie. Shamrock. Daisy. Little Old Man. Snigglers. Elvis. Li'l Richard. Jack Duggan. Noel. Peanut. The Ossabaws: Spanky, Alfalfa, Buckwheat, Stymie. Petunia. Little Rosebud. Floral. Pokémon. Sebastian. Babe. Leo. Big Rosie. Midnight. Larry. Curry. Moe. Dolly. Molly. Harrigan. Mulligan. O'Flynn. Ellie Mae. Tommy. Daisy Mae. Precious. Laverne. Shirley. Hazel. Roscoe. Phred. Rascal. Hampton. Cherie. Sergeant Major Tinker. Trooper. Oogie. Rufus. Miss Sugarfoot. Chopper Harley Hog. Wilbur. Henrietta. Big Boy. Peaches. Lucky. Finnegan. Jasmine. Bridgette. Alvin. Theodore. Simon. Dee Dee. Phoebe. Prudence. Piper. Tootsie. Bimbo. Betty Boop. Colleen. Clancy. Penelope. MacGyver. Fergie. Fleurri. Ciara. And all the others we've loved and lost. R.I.P.

Big Earl's Arrival by Laura Hoyle
Twelve–hundred pound Big Earl arrives at Pig Preserve from VA. Big Earl is Babe's first "big pig" companion.

Rich and Babe by Laura Hoyle
Babe, the Hoyle's first farm pig, was soon discovered to be a potbelly magnet.

Big Earl and Scooter by Richard Hoyle/Laura Hoyle
Big Earl having a conversation with his good friend, Scooter.

Rosebud and Rich by Anita Krajnc
Rosebud and friends enjoy afternoon peanut butter and jelly snacks.

Ciara Fiona by Caroline Wong
Ciara, with good friend, Rosebud, in the mudhole near the big barn.

Group of Pigs in Rolling Grass by Anita Krajnc
A group of pigs living life on their own terms in a rolling meadow on the forest's edge.

Silico, Feral Pig by Anita Krajnc
Silico, seen deep in the forest, enjoying the freedom to be himself.

Damien, Feral Pig by Anita Krajnc
As ferals, both Damien (pictured), and his sister (below), remain more elusive than the farm pigs.

Sunshine, Damien's Sister, Grazing by Anita Krajnc
Feral pig, Sunshine, sister of Damien, rooting and grazing in the fresh spring grass.

Jasmine with Video Camera by Anita Krajnc
The camera loves Jasmine and she's fascinated with it as well.
Great photographer too.

Big Bobby by Douglas Thron
Big "Bad" Bobby's surly attitude softened after
regular kind words, treats, and butt scratches.

Blue-Eyed Wilbur by Anita Krajnc
Having been raised alone, Blue-Eyed Wilbur was terrified of pigs when he arrived.
He became fast friends with Big Bobby, someone who stood up for him.

DeeDee Caesar Behind by Anita Krajnc
Arriving after a near death situation, DeeDee was quickly accepted by a loving pig family.

Henrietta, Part of Sunshine, Damien's Brother, Wilbur by Anita Krajnc
Sweet Henrietta was rescued from a 'farm to table' exhibit with her brother Wilbur.

Miracle and Baby Drinking from Freshwater by Tyler Johnson
Momma and Baby Miracle found Hidden Hollow's freshwater pond the day they arrived.

ANIMAL SAVE MOVEMENT GOES GLOBAL

By Anita Krajnc

1

A Promise to a Pig

Pigs are wonderful, buoyant ambassadors for animal justice. They are incredibly emotional, intelligent, curious, and fun social beings with distinctive personalities. Once you meet pigs on their own terms, you can't get enough of them. It's easy to form special bonds and friendships. To know them is to love them. When we think of the billions of pigs killed in slaughterhouses around the world each year, "ambassador" is not the first word that comes to mind. We only recognize pigs as individuals when we give them names, such as Blue Eyed Wilbur.

An ambassador has a name, a story, a character, and a community, like the pig villages at Richard Hoyle's The Pig Preserve in Tennessee. The pigs in the slaughter trucks are treated like "dumb, unfeeling livestock and property." They have serial numbers tattooed on them and specially cropped ears to identify their human "owners." The disparity between these two scenarios is the difference between absolutely magical paradise and hell on Earth.

I moved into the vicinity of a pig slaughterhouse in downtown Toronto in 2006. For four years, I did nothing. I lived less than a mile away from a slaughterhouse and could see the chimney stacks from a distance. I was already vegan and an activist, and I knew some sort of activism should center around this killing facility in our community. But for years, I said to myself, "Somebody should do something!" I figured somebody else should organize protests, and then I'd join them. I even asked activists in nearby Hamilton to start outreach events there. I pondered my neighborhood slaughterhouse only a few times a year, usually when I took a streetcar on Bathurst Street and caught a distant glimpse of the beige buildings' chimneys billowing an ominous smoke. Then, in 2010, I adopted a dog, Mr. Bean. That was the turning point and the catalyst for my creating Toronto Pig Save.

Mr. Bean and the Founding of Toronto Pig Save

In late 2010, my mom and I adopted Mr. Bean from Project Jessie, a Canada-wide animal-rescue network. Mr. Bean—a beautiful, exuberant, stubborn, very expressive, lanky beagle-whippet mix—came with that name. He also came with a strong desire for energetic walks. Each morning, we'd walk along Lake Shore Boulevard. That was the first time I saw the pig trucks, among the many trucks in rush-hour traffic. For the first time, I was directly confronted with these huge trucks transporting pigs to slaughter in the middle of my downtown neighborhood. When the traffic slowed, I could see the pigs' beautiful pink snouts poking out of the portholes. Occasionally, I'd glance at their bewildered faces full of terror and fear. The traffic would become normal once again, and the trucks would disappear, leaving only a lingering sickly smell of the unnaturally filthy inhabitants.

One day, when the traffic was especially bad, I saw eight or nine transport trucks chugging along, one after another, with pig snouts pointing out and curious, sad, and scared eyes staring pleadingly. That was it! I no longer had a choice. I had to act. I had to organize.

That was when Toronto Pig Save (TPS) was born.

At the time, I was reading biographies by Romaine Rolland, a vegetarian pacifist and writer who won the 1915 Nobel Prize in Literature and who, in the 1920s, set up a vegetarian society and organized an anti-fascist league. He also wrote biographies of Ramakrishna and Vivikananda, Tolstoy, Gandhi, and other exemplary individuals whom he considered societal role models, who, in the face of injustice, had all taken action in their local communities, despite their busy lives and careers. What impressed me was how—as accomplished people—they had dropped their regular routines and engaged in community organizing in response to egregious injustice in their communities. For example, Tolstoy and his family organized hundreds of soup kitchens in 1892 during a famine in Russia, with powerful effect.

The message was loud and clear: *It's no time to waver. It's up to you. Don't pass the buck.* Dolores Huerta, a vegetarian and co-founder of the National Farm Workers Union alongside Cesar Chavez, said: "We would say, 'First of all, nobody is going to change your conditions unless you do it. You have to do it yourselves. The next thing is that you have power. . . . You have power and the power is in your person, but it's not going to work unless you get together with other people to make it happen."[1] And there's no time like the present. As Tolstoy said in *On Life*: "Future love does not exist. Love is a present activity only. And a person who does not manifest love in the present has no love."[2]

By putting me in touch with my surrounding community, Mr. Bean had sparked the epiphany that led to the formation of Toronto Pig Save. Our mission was to unveil the terrible plight of those pigs in transport trucks and all the other animals in slaughterhouses in our city. It was time for ordinary people to organize to defend utterly innocent and powerless beings. What ensued was a David-versus-Goliath battle pitting compassionate communities against a multi-billion–dollar industry that relies on violence, killing, and . . . secrecy. We wanted to—metaphorically—give slaughterhouses glass walls through our firsthand witnessing, in the hope that this would create a wave of new vegans and activists in our community and beyond.

In December 2010, Toronto Pig Save activists met in my apartment to hold the first of a series of monthly organizing meetings, followed by

1 Cited in Eric Mann, *Playbook for Progressives: 16 Qualities of the Successful Organizer* (Boston: Beacon Press, 2011), pp. 89–90.

2 Leo Tolstoy, *On Life and Essays on Religion*, translated by Aylmer Maude. London: Oxford University Press, p. 98. *On Life* was first published in 1887, and the essays between 1894 and 1909.

a vegan potluck. Joanne O'Keefe, an animal activist from St. Catharines, Ontario, brought a sumptuous vegan raw mango pie. I mention Joanne because she would soon plan the Big Pig Trip, which I would join, with the farthest point being her dream destination—The Pig Preserve.

We decided to start with art activism, for which we would hold a series of art shows to encourage audiences to see the pigs as individuals and empathize with them. We began with an art fundraiser at Sadie's Diner and Juice Bar, followed by an *Art to Save Pigs* exhibit at the Thinking About Animals conference at Brock University in St. Catharines.

We worked with more than a dozen artists and photographers and amassed more than a hundred artworks, including prints donated by internationally renowned animal rights- and labor-conscious artist Sue Coe. Her art shows the plight of farmed animals and the conditions of slaughterhouse workers through realistic and expressionist depictions. For example, *My Sister and I Outside Slaughterhouse* portrays Sue and her sister in front of the Thorn Apple Valley slaughterhouse in Detroit. Once a Fortune 500 company, Thorn Apple Valley was then the largest facility of its kind in the US, slaughtering 13,000 pigs every day. In Sue's artwork, innocent pigs poke their snouts out of truck portholes, while the bodies of those who died during transport are strewn on the road and sidewalk in front of the sisters.

Caitlin Black, a Toronto artist and graphic novelist who regularly contributed artworks for our activities, would kindly ask us, "What do you need for upcoming campaigns?" Her artworks are another example of revolutionary, political art with a social message. In aligning ourselves with artists, Toronto Pig Save was following the advice of revolutionary composer Hans Eisler, who urged social movements to forge close contact with progressive artists in order to build new repertoires: "Find a good poet to write the history of the union, in narration and song demonstration, making sure that the concert is a lively and inventive affair."[3]

Simultaneously, our priority was to obtain footage of the activities of the local Quality Meat Packers before engaging in public protests, which

3 R. B. Davis, "Music from the Left – Part Two," *US Opera web*, Winter 2002.

would alert this slaughterhouse and cause it to take secrecy measures that would hinder us. For our undercover work, we snuck along the railway lines behind the slaughterhouse dozens of times during the early morning or late at night. Even before we drew near to the windows that opened onto the kill pens and kill floors, we could hear the poor pigs screaming at 109 decibels. In our late Sunday evening surveillance missions, we saw thousands of pigs who couldn't be killed on arrival crammed in overnight pens to be killed the next morning—a scene that Jo-Anne McArthur would later recount in her poignant, award-winning book *We Animals*.

On weekdays at 6 a.m., we witnessed gentle baby pigs being shocked and gassed. It was surreal. You could hear their agonizing, human-like screams a hundred yards from the building. At this volume, the screaming could lead to hearing loss for the slaughterhouse workers. It was hard to imagine what the pigs themselves were experiencing in those pens as they waited, sometimes for a whole day, for their turn. The pigs were screaming because they were being electrocuted with hard-wired cattle prods to race them in single file along a corridor to the carbon dioxide gas chamber. After seeing the footage we had taken, Sue Coe wrote to me: "The footage shows the pigs moving at such a fast rate into the gas chamber, being struck with the cattle prod, that visually the pigs look like they are bobbing on water. They enter the chamber, and then slip out on a slope, ready to exsanguinate. None of us, especially the majority who eat animals, should ever look away from this."

One Sunday night, we heard sounds from the kill floor that we thought came from the cleaning staff. Then, we witnessed a pig squirming in the narrow chute in front of the gas chamber doors. This pig had been left there over the weekend, unable to move, without food and water, and that was after having been violently prodded and made to witness others being forced into the gas chamber. This summarized the unimaginable callousness and systemic abuse inherent in an evil, mass-killing industry.

I felt ashamed to belong to the human species.

The elevator to the carbon dioxide gas chamber carries the pigs down into a dark basement, because carbon dioxide is heavier than air. The poison gas is supposed to render the pigs unconscious. As the elevator descends, the pigs in its meshed cage point their snouts up, trampling on one another as they try to breathe air from above to no avail. Lesley Moffat, a Canadian animal activist who works for Eyes on Animals in the Netherlands, made a documentary comparing CO_2 gas chambers with electric stunning. She believes CO_2 stunning is the cruelest method, causing the most suffering: "I believe the concentration of CO_2 at the very bottom is 80 percent. Some [animals] scream and jump, others try to get out, some just fall down. After about twelve to seventeen seconds, most are down, but not all. The longest I counted was twenty-seven seconds of one fighting to get out. It all depends on their genetics, their size, the health condition of their lungs, if they exhaled or inhaled before going into the CO_2. It is not nice. And in the law, it says that the animals must be *immediately* rendered unconscious."[4]

Lesley says one supermarket in the UK now purchases pig meat from only slaughterhouses using carbon dioxide "because there are less blood spots." That, she believes, is the real reason carbon dioxide is being used—nicer-looking pig flesh.

After the pigs have been "stunned," they are brought up in the elevator to be dropped on a counter, where a "shackler" shackles a rear leg so that each pig can be hoisted onto a conveyor belt. A "sticker" thrusts a hollow knife in the pigs' throats as they are carried along the conveyor belt so that they will bleed out. They are still alive at this point, because their hearts are needed to pump out the blood. Their bodies are then thrown into the scalding tank to loosen their hair. Plastic whips further help with this job. Torches complete it. Tracy Shepherd, a former slaughterhouse worker at Fearmans Pork in Burlington who is now a pig rescuer, says that some pigs wake up on the conveyor belt, and many are still conscious when they are thrown into the scalding tank. Butina, the Danish maker of the gas chambers, admits that 1 to 2 percent of the pigs may still be conscious when entering the tank. The true percentage is likely far higher.

It was unspeakably disturbing to see the same gentle pigs whom we had witnessed only hours before in the transport trucks now waiting in

4 Lesley Moffat, personal email communications with author. October 14, 2011.

kill pens to be tortured and murdered. Those pigs should have been at a sanctuary, such as The Pig Preserve in Tennessee, roaming the forests with their families and friends. They should not have been bred in the billions each year for human exploitation.

Pig Vigils

After spending six months building up a base, Toronto Pig Save officially launched weekly pig vigils at Quality Meat Packers in the hope of spreading the idea that we humans have a moral responsibility to bear witness to animals being transported into our hometown industrial killing facilities.

In June 2011, People for the Ethical Treatment of Animals (PETA), the world's largest animal rights group, organized a "human meat tray" protest in which activists, drenched in red paint, lay on Styrofoam in front of Quality Meat Packers. Emily Lavender, PETA's Canadian campaigner, asked if Toronto Pig Save activists could join. By garnering media awareness at the start of our campaign, PETA helped put the Toronto slaughterhouse on the map. That was the beginning of an important collaboration that included mobilizing PETA members to attend pig vigils. A few years later, they supported me as I faced criminal charges at what became known as the Pig Trial.

I covered the PETA event for RabbleTV, a progressive online Canadian news outlet where I volunteered. This involved going to a busy intersection where the transport trucks turned at a set of lights a few blocks from the slaughterhouse. On previous walks with Mr. Bean, I had seen the pigs only from a distance. Now, with the trucks stopping at the lights, I could go right up to them and gaze into a porthole.

When I got right up close, I made eye contact with a pig. Within reach, she looked up at me with pleading eyes from what appeared to be a dark and dirty dungeon. At that moment, I made a promise to that pig: "*I will help organize at least three vigils a week, for you and your friends.*" I knew that to build a movement, I had to be on the ground regularly and engage in community organizing.

This promise has been kept by Toronto Pig Save activists. Regardless of the weather, and even during special holidays, we hold three vigils every week.

Without Mr. Bean, this wouldn't have happened. We realized that bearing witness is so powerful—a life-changing experience. When we saw the sad and scared animals in the trucks going to the slaughterhouse, we knew that the world had to see them too, and that when it did . . . everything would change.

It's not enough to be behind a computer. You need to be on the ground, in the neighborhood, meeting the pigs and humans face to face to build a movement to help animals.

Animal vigils are a unique type of activism. At an animal vigil, you see the truth unfiltered. We stand in front of the gates of slaughterhouses and look into the eyes of farmed animals moments before they are to be violently murdered. At this point, they're still individuals. They haven't been disassembled yet. It's the most powerful experience I've ever had. There's an accountability that comes with being present and bearing witness. You know you have to do much more to help stop these atrocities.

The idea of holding regular vigils with the community came from the past—from Leo Tolstoy, Gandhi, and other community organizers.

My motivation to hold community vigils was based on an understanding of the importance of community organizing and nonviolent direct action. My undergraduate degree was in Peace and Conflict Studies. In his book, *Non-Violent Direct Action*, Harvard professor of Peace Studies Gene Sharp lists hundreds of methods of going about this, including demonstrations, vigils, sit-ins, and banner drops.

Multiple slaughterhouse vigils a week were needed to set off a new wave of animal activism in Toronto and beyond. Inspired by Gandhi's Salt March, Cesar Chavez and United Farm Workers' roving blockades in their union-organizing drives, and the civil rights movement's Montgomery bus boycott, sit-ins, and marches—all based on the premise of regular street actions—it was the start of a movement that would go global.

To understand Toronto Pig Save, you need to know Leo Tolstoy. Tolstoy is revered as a great author, but he was also a great activist. In *Calendar of Wisdom: Daily Thoughts to Nourish the Soul*, he provides a profound yet simple definition of bearing witness that has made its rounds around the globe through Toronto Pig Save—and later, the Animal Save Movement. Tolstoy says: "When the suffering of another creature causes you to feel pain, do not submit to the initial desire to flee from the suffering one, but on the contrary, come closer, as close as you can, to him who suffers, and try to help him."

Vladimir Chertkov, Tolstoy's best friend and confidant in later life, was, like Tolstoy, an ethical vegetarian. In his 1912 book *One Life*, Chertkov animates the practice of bearing witness and the obligation to do so: "A simple reminder around a dinner table that a meal being served consists of dead animal parts tends to kill the appetite and makes the diners indignant. Nothing more significantly reveals the disgusting and illegal nature of this action than the need to hide its true meaning from oneself. To get a true notion of this matter one, first of all, has to face it. And the best way to literally 'face' it is by visiting a slaughterhouse or a kitchen yard and firsthand witnessing the killing of livestock or poultry for our table. I have no doubt that the great majority of people who would do it several times with diligence very soon would recognize the unlawfulness of what is happening before their eyes."[5]

In one of the earliest of such records, Leo Tolstoy went *inside* a Russian slaughterhouse to purposely bear witness in 1891. He was determined to see with his own eyes the reality of how animals were killed at the slaughterhouse as part of his research for an introductory essay for Howard Williams's *The Ethics of Diet*, a book of short essays on sixty-nine famous vegetarians in history, including Pythagoras and British poet Percy Shelley. Tolstoy called his essay "The First Step" because he felt that the path to a good life is like a staircase and one has to begin with vegetarianism as the first of many steps in one's moral progress.

5 Vladimir Chertkov, *One Life: Concerning the Killing of Living Beings*, translated by Anya Yushchenko, 1912.

His ideas were premised on the belief that people are inherently good and that "deep-seated in the human heart there is an injunction not to take life!" Flesh eating," he argues, "is simply immoral, as it involves the performance of an act, which is contrary to moral feeling—killing."[6]

In the essay, Tolstoy presents in graphic detail the horrific killing methods he witnessed. It was worse than he had imagined: "I fancied that, as is often the case, reality would very likely produce upon me a weaker impression than the imagination. But in this I was mistaken."[7]

In the account, Tolstoy first notices the smell, which the animals must experience too, only at heightened levels, as cows can smell blood from miles away: "Even at the entrance one noticed the heavy, disgusting, fetid smell, as of carpenters' glue, or paint on glue. The nearer we approached, the stronger the smell became." The terrified animals struggle with the butchers, who do not notice them and are unconcerned with their extreme suffering. Tolstoy witnesses animals stabbed and bludgeoned, one by one, before having their throats slit and their legs cut off while they are still conscious.

> Through the door opposite the one at which I was standing, a big, red, well-fed ox was led in. Two men were dragging him and hardly had he entered when I saw a butcher raise a knife above his neck and stab him. The ox, as if all four legs had suddenly given way, fell heavily upon his belly, immediately turned over on one side, and the butcher began to work his legs and all his hind-quarters. Another butcher at once threw himself upon the ox from the side opposite to the twitching legs, caught his horns and twitched his head down to the ground, while another butcher cut his throat with a knife. From beneath the head there flowed a stream of blackish-red blood, which a besmeared boy caught in a tin basin. All the time this was going on the ox kept incessantly twitching his head as if trying to get up, and waved his four legs in the air. The basin was quickly filling, but the ox still lived

6 Leo Tolstoy, "The First Step," in Howard Williams, *The Ethics of Diet: An Anthology of Vegetarian Thought*. Guildford: White Crow, 2009.

7 Tolstoy, "The First Step."

> and, his stomach heaving heavily, both hind and fore legs worked so violently that the butchers held aloof. When one basin was full, the boy carried him away on his head to the albumen factory, while another boy placed a fresh basin, which also soon began to fill up. But still the ox heaved his body and worked his hind legs.

The same poor ox takes minutes to die as he is skinned and dismembered while alive and conscious. It reads like a modern-day exposé of slaughterhouse dungeons. R. F. Christian writes: "Tolstoy's horrific details of the repeated futile attempts of tortured, wounded and partially eviscerated animals to escape their violent end helped convert a number of people to the vegetarian cause."[8]

Pig Island

More than a century later, in July 2011, we began our weekly pig vigils. We had a heart-wrenching introduction to the horrors going on in our backyards that few had paid attention to before.

Each week, two vigils were held on weekday mornings at a traffic island a kilometer away from the slaughterhouse that the trucks passed, often stopping at the red lights. The early-morning pig vigils usually started at seven or eight and lasted three hours. In our first year, our modest initial group of a handful of people graduated to five to ten regular vigil attendees, mostly women, bearing witness on a traffic island that we dubbed "Pig Island."

During rush hour, the transport trucks would become stuck in traffic, forcing them to stop longer at the lights. At times, there would be trucks with pigs too despondent and frightened to approach us through the portholes. More often, pigs would come to us and nuzzle our fingers, greeting us with friendly and curious grunts. The most heart-wrenching instances were when the pigs, desperate and confused, looked out at us with imploring eyes, seemingly asking us what was happening and wondering if we would be the ones to free them. They had known nothing

8 R. F. Christian, "Tolstoy and The First Step," *Scottish Slavonic Review*, 20, 1993.

but abuse at the hands of humans, yet they still reached out in hope and friendship. We would touch the snouts of these baby pigs—usually just four to six months old. As the transport trucks pulled away, we would say, "We love you," and, "We're sorry."

Pig Island was an ideal location. This busy intersection in downtown Toronto allowed us to easily and safely bear witness up close when the lights turned red, while alerting thousands of drivers and passersby to our vigils and the messages on our banners and placards. As activists tried to reach the public, drivers reached back with a cacophony of honks. We'd hand out leaflets and sometimes offer vegan BLTs to the drivers. Between trucks, we would talk among ourselves about organizing.

Bearing witness politicizes the transport trucks and gives visibility to the pigs when we invite the public to look and join us. The animal exploitation industry depends on subterfuge to hide suffering. Before starting Toronto Pig Save, I had never even noticed the trucks, even though I was a vegan and an animal rights activist. Even these transport trucks were hidden in plain sight. Toronto Pig Save helped highlight the massive scale of the routing killing of animals for human consumption and, simultaneously, each animal's individuality. The images of the animals in death trucks crying out, "Face us! Help!" brings to light the incalculable pain and horror of each individual animal who wants no more than to escape confinement, torture, and death, thus breaking the disconnect maintained by cellophane- and plastic-wrapped meat, dairy, and eggs. Our mission is to show that those in the transport trucks are individuals, not property. Each and every pig in a slaughter truck wants to live, to be happy, and to be free, just like every one of us. All pigs deserve our empathy and our commitment to help.

On Sunday mornings, a third vigil was held, this time at the front gates of the Quality Meat Packers slaughterhouse. Activists gathered on the sidewalk in front of the long drive leading to the pig-unloading area. As trucks waited in line to be unloaded, activists stood and listened in horror to the universal language of suffering, the pitiful groans of hot, dehydrated, overcrowded, wounded, exhausted pigs at the end of a long and arduous journey to the slaughterhouse. They let out piercing screams as the drivers stuck electric prods in the portholes, sometimes hitting them in the face with these instruments of torture.

Across the street from this hellish scene of pigs in the heat waiting to die was a residential area with a dog park, where dogs frolicked by a water fountain, playing with each other and digging, just as pigs like to do. Meanwhile, trucks packed with distressed pigs would drive by.

Toronto, like Chicago, used to be called "Hog Town." The slaughterhouse had been there, next to the railway tracks, since the 1800s. But in recent times, a Liberty Village condominium development emerged nearby, which was making the slaughter truck traffic untenable in the yuppy village.

We are often asked: *"Does it help to be present and bear witness, knowing you can't do anything for those specific pigs doomed to die?"*

Our form of bearing witness is only *partial*. To fully bear witness would mean helping the pigs in the transport trucks in the same way we would want others to help us if we were in those circumstances. While there are paradoxical and tragic elements to our philosophy and approach to bearing witness, being present to show love and empathy, even if only for a brief moment, is better than looking away. We who bear witness have a responsibility to not look away when an injustice is happening—to try to help the best we can. In addition to holding vigils for those animals who can't make it, we strive to raise awareness through collective public protest, in the hope of preventing other pigs from being raised for slaughter. Our hope is to build enough momentum and a movement that will eventually put a stop to all slaughter—to be able to say to the exploited and suffering pigs before us: *"We will do our utmost so as not to let you die in vain."*

The pigs know whose side we are on! A transport truck driver once warned us not to put our hands near the pigs as they'd bite us. Nothing of the sort ever happened. The gentle pigs knew we were their friends.

Another time when I got the sense that the pigs knew I was on their side was when I was on the slaughterhouse premises, right in front of the unloading dock, on a bitter winter day. After I crossed into the company's property to confront the managers about the hypothermia and frostbite suffered by the pigs in the transport trucks, along with the gross injustice of their impending slaughter, the pigs came towards me as I stood between them and the managers and raised my voice in their defense.

They know.

The collective act of bearing witness helps build community by empowering activists, the public, and workers in the animal exploitation industry to care and speak out. It sends out the message: "We're here, we care, we're sorry, and we're trying." We plant some seeds, and we put the issue on the public agenda. Suddenly, people start seeing the transport trucks and peaceful vigils advocating for the pig victims in the trucks, then talk about them at home, at school, and at work.

Bearing witness affects the witnesses the hardest. By virtue of being physically present to bear witness, formerly non-political people become activists. Activists deepen their commitment by leaps and bounds, with many saying, "I need to be there each week," and many making animal liberation a lifelong priority. Our aim is to bear witness at every slaughterhouse and build momentum to shut down these killing facilities.

Bearing witness is essentially an ethical strategy to engage one's own and others' empathy in considering the plight of pigs. As an onlooker, one absorbs the forlorn souls mirrored in the pigs' eyes. It is painful to go to a slaughterhouse and be present, but social change requires sacrifice. Paradoxically, it is also empowering. As nineteenth-century Indian prophet Ramakrishna stated: "But my heart has grown much, much larger, and I have learnt to feel [the suffering of others]."[9]

How will the world change for the majority of pigs unless we put our feet in their hooves? We all need to have a bigger heart for our animal

9 Romain Rolland, Prophets of the New India, translated by E. F. Malcolm-Smith. New York: Albert & Charles Boni, 1930, p. 314.

friends—and it's not hard to do. Animals have the biggest hearts of all. We just need to emulate them.

We hoped that our weekly vigils would change the community's obliviousness. From 2011 to 2013, Toronto Pig Save held its annual Labor Day "veggie dog giveaways" across the street from the pig slaughterhouse, inviting local residents to enjoy live music, free veggie dogs, face painting, chalk art, and information tables and videos. Many local residents who joined us did tell us that they changed their eating habits after seeing the pig trucks or hearing the pigs' screams and reading our banners and placards.

Also balanced against the difficulty of bearing witness are the benefits that community organizing brings to activists. Many friendships have arisen from our animal vigils and other events. We know that it's important to take good care of each other. Everyone is a leader, and good leadership is defined as bringing out the best in others. Community organizing is, by its very nature, inclusive, and community organizing using a Gandhian, love-based approach is contagious.

New activists bring new energy and ideas to the animal vigils. Sharmini Gana, who joined us in early fall of 2011, would shout, "Go vegan!" with a big smile to drivers, including those at the wheels of transport trucks. When the transports rolled to a stop next to Pig Island, she would clasp her hands in prayer, imploring them to take leaflets. Before she joined us, we had never had any success engaging the pig transport drivers. Sharmini's presence also brought out more honks of support from ordinary commuters.

Some of our members began as omnivores and shifted to a vegan lifestyle as a result of bearing witness. For example, Sylvia Fraser, author of *My Father's House: A Memoir of Incest and of Healing*, was cycling by one winter day when she noticed three of us standing in front of Quality Meat Packers with placards that read HAVE COMPASSION and 6,000 PIGS KILLED EACH DAY. As a local resident, she had been offended by the odious stench emanating from the slaughterhouse, especially when the blood vats were being emptied into blood trucks. After reading our placards, she realized in an instant that the smell was the only way the pigs could protest their untimely and cruel deaths. She got off her bike and picked up a placard. At that time, she thought nothing of eating a ham sandwich for lunch. Now, she is vegan.

Unforgettable Portraits

Further afield, the vigils have impacted scores of people who view them on social media and through mass media coverage. The weekly stream of photos and videos of the pigs allows a much wider audience to vicariously witness animal exploitation and suffering, to listen as the animals tell their stories, to see them as individuals, and finally, to observe the power of collective action. We encourage all attendees to bring cameras or use their mobile phones to capture their own photos and videos at the vigils and post them on Facebook, Instagram, YouTube, TikTok, and other social media platforms, using hashtags and tagging friends. This way, we have been able to reach vast networks of people who would not have otherwise been exposed to this message or paid attention to this new form of activism, were it not presented by someone familiar. Countless photographers and videographers have attended vigils, each documenting the many animals sent to slaughterhouses in their own community. This has resulted in substantial documentation and profound images and stories of farmed animals being placed into the public record.

The most viral footage of Animal Save Movement, with 5–20 million views on Facebook under the title "Best Video You Will Ever See," is of activists giving water to thirsty pigs. Toronto Pig Save and Los Angeles Animal Save's pig vigils have produced videos fraught with palpable emotions from the pigs and activists. Viewers feel as though they are present as tearful activists try to help the desperate pigs by delivering water through the holes of the trucks' metal walls.

A 2015 Faunalytics study, *Creating Change Online: A Study of The Save Movement's Social Media Presence*, found our audience to be 59 percent vegan, 13 percent vegetarian, and 26 percent omnivore.[10] While these figures suggested our groups could be mostly preaching to the choir,

10 Faunalytics, *Creating Change Online: A Study of The Save Movement's Social Media Presence*, 2015.

Faunalytics concluded that we had played a significant role in building that base and helping to retain vegans as well as fuel their activism. Compared to other available information, Animal Save Movement's social media posts give 70 percent of our omnivore followers a greater reason to consider eating fewer or no animal products in the future, and 80 percent of vegetarians and vegans who follow us a greater reason to maintain their diets. Ninety percent of our supporters are advocating for animals online or in person, and 78 percent of those credit our footage for inspiring their activism. Over 90 percent of supporters find Animal Save Movement's video and photo content impactful.

Slaughterhouse Workers

Bearing witness has myriad effects on slaughterhouse workers. Since we began our pig vigils, the workers have become less violent and cruel in their "handling" of the pigs. While unloading them from the trucks, workers are no longer verbally abusive and make less use of electric prods. This brings to mind what Romain Rolland wrote in his book *Popular Theater*: "The evil that is seen face to face, the evil that is conscious of being seen, is more than half conquered."[11] During the weekly vigils, our group members talk with passersby as well as slaughterhouse workers, owners, and managers, using a love-based approach inspired by Tolstoy's method of connecting the dots for those immersed in evil institutions and practices. Tolstoy writes of the necessity for activists to talk to their opponents and point out wrongdoing: "But, however much [the wrongdoers] try to deceive themselves and others, they all know that what they are doing is opposed to all the beliefs which they profess, and in the depths of their souls, when they are left alone with their conscience, they are ashamed and miserable at the recollection of it, especially if the baseness of their action has been pointed out to them."[12]

---

11 Romain Rolland, Romain. The People's Theater, translated from the French by Barrett H. Clark, New York: Henry Holt and Company, 1918.

12 Leo Tolstoy, *The Kingdom of God Is Within You: Christianity Not as a Mystic Religion but as a New Theory of Life*, translated by Constance Garnett. New York: 1894.

It is important for ordinary people to stand up to the violence in their communities, as artist Sue Coe puts it: "Witnessing and standing on the kill floor changes the dynamic of everything. . . . You don't need money, you don't need credentials. Just placing yourself in the vortex creates change. That's the beginning of change."

Karen Davis, a veteran animal rights advocate and founder of United Poultry Concerns, similarly points out: "The number one thing is you need to tell people what you're about. It will sort out confusion. Be truthful and authentic. People respect you for being upfront in Virginia in a chicken factory farming area. We wear our gear and pins."[13]

At the same time, we have no illusions. The Quality Meat Packers slaughterhouse kills six thousand pigs each day—extreme violence is inherent in its operations. Inside the facility, to keep up with the speed of the "production line," workers are forced to use hard-wired electric prods to coerce the pigs down the single chute towards the carbon dioxide chamber at a rate of around eight hundred to a thousand pigs per hour!

It's apropos to recall what Charles Dickens said after conducting research for his novel *Nicholas Nickelby,* for which he visited many abusive boarding schools where boys were famished, ill, and blinded due to negligence. His response was that "these institutions had to be destroyed." Just imagine: no more slaughterhouses, no animal research labs, or other centers of animal exploitation defacing the Earth!

Quality Meat Packers Goes Bankrupt

Quality Meat Packers in downtown Toronto was the second-largest pig slaughterhouse in Ontario. Every day, about thirty transport trucks, each carrying two hundred or so pigs crammed into three levels, would travel along the city's busy throughways. In April 2014, the slaughterhouse closed before declaring bankruptcy a month later. The last transport trucks were seen on April 3, 2014, at our weekday vigil. Newspaper reports attributed the closing mainly to the rise in the price of pigs. Three years' worth of

---

13 Personal communication with Ian Purdy, March 2015.

protests, which were building in size, no doubt played a part too, though these were not acknowledged by bankruptcy trustee Farber.

The vigils expanded public dialogue about the slaughterhouse by focusing on the pigs and their suffering and on the need for community action as central concerns. Previously, the predominant discourse in the community, as represented by news coverage at least, had been limited to issues of deflated property values and the smell emanating from the industrial killing facility that destroyed the lives of six thousand pigs a day. Our community-organizing approach was based on undertaking regular, intensive, on-the-ground events—as practiced by Gandhi, Cesar Chavez, King, and others. The basic idea of holding regular weekly vigils, including on holidays, was critical to the growth of Toronto Pig Save and to the subsequent rise of similar chapters worldwide. Three vigils a week would amount to 660 vigils over the course of four years.

Community organizing is defined as the mobilization of people to take action at a particular site in response to injustice within, or in close proximity to, the area where they live. The site of community organizing may be a specific neighborhood; an industrial facility, like a slaughterhouse; or a college campus, where students are eager to participate in new movements for social change. Such a site may also be unconventional and mobile, like a transportation route of slaughterhouse trucks. Community organizing may be based on identity related to class, gender, or sexual orientation, or based on commitment to animal rights or the environment, for example.[14]

With the closing of the Toronto pig slaughterhouse, it was difficult for us to maintain our number of weekly vigils. Instead, we shifted our focus to the cow and chicken slaughterhouses in Toronto and joined activists in Burlington, a nearby city outside Toronto, which had a pig slaughterhouse. Ontario's largest slaughterhouse, Fearmans Pork Inc. slaughterhouse, owned by Sofina Foods, killed ten thousand pigs daily. We were prepared for the move to join the Burlington animal activists.

---

14 Guy Delgado, *Beyond the Politics of Place: New Directions in Community Organizing in the 1990s*. Oakland, CA: Chardon Press, 1999.

The approach of holding regular activist events is used by a number of global movements. For example, for three weeks starting in September 2018, Greta Thunberg and her fellow students staged daily school strikes in front of the Swedish Parliament ahead of the national election. Then, they continued weekly school strikes on Fridays, called Fridays For Future, which subsequently inspired several global school strikes each year. In the same vein, Toronto Pig Save started in July 2011 with three vigils a week until Quality Meat Packers went bankrupt. At the height of the movement, our attendance had increased to about thirty people at each of the weekly vigils in the Toronto area, more at special, all-day vigils. The frequent, regular vigils helped build the organization and also spread the movement in Canada and beyond.

2

The Big Pig Trip

THE FIRST OF MY THREE visits to The Pig Preserve was in May 2013, when Joanne O'Keefe organized a multiple-sanctuary road trip dubbed the "Big Pig Trip." Joanne had worked wonders promoting and fundraising for The Pig Preserve; now, she wanted to meet Richard and Laura Hoyle and all the pigs in person. Toronto Pig Save activist Caroline Wong and I had the privilege of joining her.

We started with Farm Sanctuary in upstate New York. The first time I stepped onto its pastures, I felt like I was on another planet. It was paradise on Earth. If only more people could visit farmed-animal sanctuaries and get to know these precious, kind, and loving pigs, cows, sheep, goats, horses, chickens, turkeys, and fishes when they are free and loved in forever homes—then everyone would be happier.

From there, we drove to Woodstock Farm Animal Sanctuary, also in New York State, and stayed at its delightful bed and breakfast. There, I filmed the extremely noisy and impatient pigs as they roared eagerly, demanding their caregiver speed up his preparation of their food. At nearby Catskill Animal Sanctuary, we met its lovely founder, Kathy Stevens, who told us about their school outreach program and vegan cuisine course. At Indraloka Animal Sanctuary in Pennsylvania, we were introduced to the turkeys, who wandered freely and liked eating an assortment of fruits. One turkey enthusiastically devoured a mango, the result being sticky mango getting all over her face and white feathers.

After that, we spent a few wondrous days at The Pig Preserve as welcome guests of Richard and Laura Hoyle, who prioritized the wellbeing of the dogs, cats, and pigs in their care before their own. As Richard joked: “It is definitely a zoo here most of the time! The house is always eventful from dogs, cats, pigs, and humans wandering through at all hours of the day and night, but the animal children are all well taken care of, and dearly loved, and respected as individuals.”

The Pig Preserve, where one hundred pigs foraged over one hundred acres of natural habitat in the beautiful, hilly forests of Tennessee, was so different from the previous farm sanctuaries. As Joanne exclaimed merrily: “The pigs are not in little paddocks—they have places to roam!”

The pigs were so comfortable because they were in their own space. We were visiting their home. I enjoyed following Twiggy—part feral and part Berkshire—on the hilly trails. She was a former show pig who was now enjoying life in a natural environment, no longer serving anyone.

I fell in love with Dee Dee.

As you may remember, she had been hung upside down by a man who tried to butcher her alive, until a neighbor intervened with her shotgun. Since Dee Dee’s legs had been damaged from being strung up, she screamed in pain when Richard and I tried to move her into the barn for the night. Richard said that if we waited patiently, she would signal when she was ready to try again, and she did. Dee Dee knew we were there to help, and once inside, she was content. When I fell into the mud midway with my camera in tow, Richard said: “Now you’re a pig person. No one is a real pig person until they fall into the mud.”

It was so nice to meet Rosebud, a highly sensitive and emotionally intelligent pig whose feelings were easily hurt. After Richard bopped her on the butt for insisting on more food, she ran up the hill into the forest, then wouldn’t return until Richard apologized by bringing her some apples. Later in our visit, Rosebud came to the rescue when Joanne became the unwilling "playmate" of a young, energetic, three-hundred–pound pig. As befitting a matriarch, Rosebud regally sent him away with a low, quiet grunt, then slowly returned to her peaceful afternoon nap.

Caroline Wong said that while at the Preserve, she connected with pigs as a species. She gave them close attention by grooming them, removing ticks, and providing belly rubs. Before the trip, she had found pigs big and a bit scary, but she soon discovered them to be gentle giants, incredibly playful and fun, and full of love and kindness.

Jasmine had fallen off a pickup truck in North Carolina, severely battered and bruised, when she was only a week old. Richard had bottle-fed her for the first few days, and she had slept in his bed, snuggled up under his shoulder. As he told us: "When she was hungry, she would get up and suck on my ear. I would sleepily roll over and grab a baby bottle of pig formula off the nightstand. She would suck on it until she was full, then she would burp happily and go back to sleep. This went on until we got her weaned."

Jasmine followed us everywhere, often rubbing against my tripod and video camera. It was also during our visit that she ventured into the forest for the first time to feed with the other pigs. As Richard said: "She's growing up. She's more of a pig!"

Caroline was struck by the pigs' "human-like" behavior: "They argue with one another. Some don't want to eat together. They connect with some pigs more than with others and have best friends. Sometimes, they get upset and won't come out for dinner." Peggie Sue didn't like eating with other pigs, so she carried her bowl off to eat alone and in peace. "Most people are unaware of how sentient and how aware pigs are," continued Caroline. "If people got to know pigs, they would realize they are not food."

After our big trip to The Pig Preserve and the other sanctuaries, Joanne, Caroline, and I found it more heartbreaking to bear witness at the pig vigils. Now, we knew, from personal experience, that each of the pigs looking at us through the portholes of a transport truck, on their way to be tortured and slaughtered, might have been innocent Jasmine, or sensitive Rosebud, or long-suffering Dee Dee. Now, we knew that everyone who ate pig flesh was eating one of our friends. The experience also made any animal rescues from slaughter trucks or slaughterhouses more poignant, as we now knew the possibilities of what could be.

Animal Rescues

At our Toronto Pig Save vigils, we often thought of rescuing pigs. Once, we heard of a pig who had escaped briefly in the driveway. City officials, who owned a property next to the slaughterhouse, said that they had helped herd the pig back to the unloading dock. We weren't there at the time, but our hearts ached for the bright-eyed pig who had hoped to avoid the horror that was the kill pen and kill floor. We never had any relationship with the slaughterhouse owners other than their calling the police on us when we occasionally trespassed to speak up for the abused pigs.

In Australia, Perth Pig Save organizers had a much better relationship with the Linley Valley Pork slaughterhouse in Wooroloo. They had reached an agreement with the slaughterhouse whereby the trucks would stop for two minutes so activists could safely bear witness. At Christmas in 2017, activists even managed to rescue a sow. As it happened, an activist approached the general manager to ask if he'd spare a life before Christmas. He came back within half an hour and agreed. The sow was named Carol in the spirit of Christmas. As organizer Jen Regan recalls, Perth Pig Save saw its attendance tripled at that time.

Carol would now live out her life at Greener Pastures Animal Sanctuary after having known only confinement. Her muscles were weak from her having been trapped in a sow stall for most of her life. When Carol arrived at the sanctuary, she still had the marking that would have sentenced her to slaughter—pink spray paint on her back—but it didn't take her long to enjoy freedom. That same day, after a little warming up, she got excited and started running and dancing around the paddock happily. She also had her very first mud bath. Jen says: "Now, a few months on, Carol has settled well into her new sanctuary life. She was introduced to the other pig residents, has established herself within the pecking order, and has seemingly even adopted a son, Iggle Piggle, a younger pig. The two are inseparable and are often found cuddling together. We like to think of Iggle Piggle as the son she never got to keep, having had between 80 and 120 piglets taken from her in her four-to-five–year lifespan." Carol is now thriving with her new family.

Simply by being present at slaughterhouses, activists are able to rescue many animals who escape from transport trucks while being unloaded. For example, in April 2018, activists successfully negotiated the release of two mother sheep and their two babies from Newmarket Meat Packers in Ontario as an act of mercy just before Easter and Passover. While activists were there to pick up the lucky four, another baby lamb escaped from a transport truck. While the truck driver was sure that this lamb would come back to his group (then eventually be slaughtered), the lamb ran into the nearby forest and disappeared. Hours later, activists found him (whom they named James after James O'Toole, one of the activists on site) and took him to Dog Tales Rescue and Sanctuary to join the other freed sheep. A couple of weeks later, at a Las Vegas Animal Save vigil at a bird slaughterhouse, Giselle, an organizer, saw a duck escape from the unloading area. As she picked up the duck, a worker ran to her, grabbed her hair, and violently pulled the duck's wing. Giselle refused to let go and adroitly rescued the duck.

According to Bobby Sudd, the lead organizer of LA Chicken Save, who holds chicken vigils every other Thursday from three to five in the morning: "There are times when a driver will give us a chicken, but that happens less since COVID-19. If a chicken arrives and has escaped the crates on their own, or if the chicken is out of reach, we will ask the employee if we can take them. We will rescue chickens as long as I have confirmed that they have safe forever homes to go to." At first, activists would agree to foster rescued chickens, but oftentimes, a chicken would turn out to be a rooster and building managers would demand their tenants get rid of the chicken if they didn't want to be kicked out. Now, instead, the chickens go to sanctuaries like Kindred Spirits Care Farm (which took in the first rescued chickens), Little Bitty Animal Sanctuary, and Steampunk Rescue Farm. Their stories are, of course, shared widely on social media.

Julia was rescued in 2021. She was pulled off a truck and taken in by a local sanctuary. Over the next few months, additional chickens were rescued and accepted by the same sanctuary, and Julia became mother hen to all of them. At another LA chicken vigil, Bobby was setting up when a truck arrived early. When they pulled the first crate of chickens off, they saw that there was one chicken who had escaped, but her foot had been crushed underneath the crate that had just been removed. Bobby grabbed the chicken, whom he named JB, and took her to a sanctuary, which got her to a veterinary surgeon that same morning. She had to have one and a half toes amputated but fortunately made a full recovery quickly. JB stayed at Bobby's place for the weekend, until the sanctuary was ready for her. Bobby recounts: "Like nearly all of the rescued chickens, she was scared at first but quickly became a cuddle-bug. I would take her outside to be in the grass for the first time, and I would watch her rediscover her nature as a chicken, scratching, digging, and dust bathing. When a garbage truck drove by she got scared and jumped right into my arms."

Special vigils have been quickly organized at sites of slaughter truck accidents and rollovers. In their desperate attempt to rescue animals, activists often battle police, government officials, and slaughterhouse managers who prefer to march the surviving animals to the kill floor.

On October 5, 2016, a horrific pig truck rollover occurred right in front of Fearmans Pork slaughterhouse. We mobilized activists to go to the location and appeal to police, firefighters, slaughterhouse management, and government officials on the scene to release the pigs. Toronto Pig Save organizer James O'Toole recalls his impression upon arrival: "Nothing could have prepared me for what I was about to witness. The second I opened my car door, screams filled the air. Relentless, prolonged, agonizing screams." James and other activists livestreamed for several hours. The slaughterhouse workers, not the firefighters, were the first to enter the truck and immediately started hitting the injured pigs to force them to leave the truck promptly. Lori Croonen, Toronto Pig Save organizer, says, "I saw hope in the pigs' eyes when they touched the grass and started rooting. They thought they were finally free. That was the saddest part for me." James adds: "Protesters were remonstrating with police, dismayed at their handling of the situation; they were clearly helping Fearmans cover up the truth. They were prohibiting protesters from approaching the truck by putting up a yellow caution-tape barrier but allowing Fearmans slaughterhouse employees past the barrier so they could hold up cardboard to shield the truth."

Steve Jenkins of Happily Ever Esther was on site with a wagon, pleading for the release of two injured pigs to his sanctuary, their would-be new home. As a few of the pigs were crossing Fearmans's parking lot to a grassy patch, we witnessed one of the pigs walk over to another pig who was injured. We named them Bonnie and Clyde. Clyde nuzzled Bonnie, trying to help her. As James remembers: "Sadness consumed me. The fact the pig was up and walking meant he could be 'processed.' Sure enough, this pig was soon led away. Bonnie was far too injured to walk and needed urgent medical attention. We were ready, willing, and able to assist to take her to the best veterinary care available." Instead, Bonnie was covered with cardboard and shot in the head with a bolt gun meters away from about a dozen animal activists crying out for mercy.

Instead of being given medical assistance and handed over to an animal sanctuary, all the surviving pigs were either shot in the head with a bolt gun or marched to the slaughterhouse, where they were electrically prodded to a carbon dioxide gas chamber. I was charged again—this time for obstructing a peace officer—for crossing the police line in an attempt to take photos of the pigs, although Fearmans's office workers were behind the police line, holding cardboard paper up high to hide the crash victims. Nevertheless, the charges of obstructing police were subsequently dropped.[15] The rollover revealed the complete moral

15 Liam Casey, "Pigs Marched to Slaughterhouse after Truck Overturns in Burlington, Ontario," *Globe & Mail*, October 5, 2017; Hannah Sentenac, "Pigs Injured in Canadian Pork Truck Accident Slaughtered, Then Thrown in Dumpsters Despite

bankruptcy of the animal agriculture industry, as Lori phrases it: "There was a complete lack of empathy. It showed the sociopathy of the system where there's zero empathy for what the pigs inside the trucks were suffering. It's institutionalized sadism backed by the government."

Valezca Lango Munsuri, lead organizer of Mexico City Animal Save, tells me of their pig rescue efforts in Mexico City almost five years later, on February 7, 2020, after a truck rolled over about ten minutes from the slaughterhouse. Valezca was alerted on social media and was informed that two activists were at the crash scene asking for help. There were about two hundred pigs in a double trailer with three floors. She sent a message to the activists and coordinated with the two of them. Valezca went to the scene herself to try to help the pigs, livestreaming the situation so as to call on other activists to join.

When she arrived, she saw the two activists bravely lying in front of the truck, trying to prevent the pigs from being taken to the slaughterhouse. Valezca talked with the authorities, begging them to liberate the pigs. According to her, "in Mexico, when something like this happens, it's illegal to consume [the animals'] bodies." The authorities agreed with her and said they would try to help. In the meantime, activists gave water to the pigs, most of whom were injured or dying.

Valezca then noticed that the slaughterhouse workers were dragging some pigs by the ears and kicking them onto a truck. Activists immediately tried to separate the pigs from the workers, telling the workers that the pigs were suffering and shouldn't be treated like that. Valezca recalls: "By this time, we were furious and defended the pigs with all of our strength. I remember I never cried at a vigil and I always got myself straight and focused, but this time, I lost my nerve and I started to cry and yell at them that pigs have feelings and they deserve respect. We started to move them into a makeshift corral on the sidewalk. The pigs were so smart, they didn't need to be hit. They understood pretty well that they needed to go to this safe area."

Offers for Sanctuary," *Miami New Times*, October 27, 2016.

After an hour, the authorities wanted to "clean up the scene" so they could reopen the freeway. They turned the trailer, with the many pigs still inside, upright. The pigs with open wounds screamed and bled as this maneuver only hurt them more. Next, additional female police officers arrived on the scene and started to aggressively push the animal rights activists away from the pigs, while additional slaughterhouse personnel arrived with knives in their hands. Valezca says: "We immediately started to resist because we knew they would start to kill them. Fortunately, more activists started to arrive. There were about three dozen activists ready to fight to save the pigs. Also, we received media attention, so they didn't dare to slaughter them on the site. (This happens often in these rollover situations, at least in Mexico.)"

Valezca continues: "By evening, the authorities left and the pigs' 'owners' arrived and wanted to take away the pigs. We started to lie down in front of the truck, but the cops started to push us away. They arrested several activists and took them into custody in the patrols. I tried to calm everyone because we were mainly women and the freeway was so dangerous for us. Sadly, they took the truck with the majority of the pigs. But I noticed that sixteen pigs were still on the sidewalk, so all the activists surrounded them. We already had a truck, which an activist had arranged, waiting in the wings. We started to put the pigs in our truck, but the owners arrived again—this time with two men with machine guns! They sat in front of our truck and started threatening us. Again, I started to mediate and try to de-escalate the situation. A female officer arrived at the scene and arrested these men, so finally, at 1 a.m., we started to put the pigs in our truck and take them to a safe place."

Valezca says what followed were many difficult months. Activists had to learn how to take care of the pigs along the way. Several pigs would not see the end of the journey but died of heat and injuries. Ten survived. Four of them were placed in adoptive homes, and six went to Granjita TyH, Huerta Vida Digna, and Santuario Eli animal sanctuaries. As Valezca reflects on the experience: "This rescue was made possible because of the unity of three dozen animal rights activists who put themselves in danger to rescue the pigs and have put in all the effort since then to take care of them. This shows how powerful a group of committed people can be. We had some experience with the slaughterhouse vigils, but nothing prepared us for all that we witnessed and for all that we had to do to take care of those sweet, innocent rescued pigs."

3

Turning the Tables at the Pig Trial

On June 22, 2015, an incident occurred that changed the course of Toronto Pig Save and propelled the act of bearing witness to suffering pigs far and wide, even more so when videos and photos taken by our activists were aired in mainstream media.

I was participating in one of our usual pig vigils at Fearmans slaughterhouse in Burlington, Ontario, alongside Frances, her son, and Nikki. It was a scorching-hot day. Nikki and I crossed at the lights and went onto the traffic median to bear witness while the other activists remained on the sidewalk holding placards. As a truck stopped at the red light near us, we could see pigs inside who were hot, thirsty, and covered in scratches. They were panting at a fast pace and foaming at the mouth. I said, "Let's give water to the pigs." As I got closer to the side of the truck, a desperate pig immediately came up to me, then reached her snout out of the truck's porthole to drink water from my bottle.

Jeff, the truck driver, jumped out of his cab and shouted, "Don't give them water!"

"Show some compassion," I replied, as Nikki, my fast-thinking companion, began capturing the encounter on her camera.

The driver demanded that I stop.

I quoted the Bible, Matthew, Chapter 25: "Jesus said, 'If they are thirsty, give them water.'"

The driver continued to shout: "They aren't human, you dumb frickin' broad! Stop giving them water."

What mattered to me was the thirst of this individual pig, not the driver's demands that went against the Golden Rule. I put my water bottle back against the porthole, following my conscience and a universal and unbreakable tenet that required one to treat others as they'd like to be treated. The same pig eagerly approached me for more water, despite the commotion. She knew we were at this vigil for her.

The driver yelled, "I'll hit the bottle out of your hand if you don't stop."

I told him that that would constitute assault.

He replied, "I'm going to call the police."

"Call Jesus," I retorted, thinking he might be a Mennonite. Mennonites are pacifists but have not yet extended their beliefs to the war on animals.

At that, the driver hopped into his truck and drove off.

Two months later, a constable showed up at my door. I was issued a summons charging me with criminal mischief, defined as interference with property, for giving water to that thirsty pig. It was a charge that carried a potential sentence of ten years in prison and a $5,000 fine. I was surprised, since for two years, we had been giving water to pigs at vigils, often in the presence of police, who on occasion even sympathized with our small acts of mercy.

It was heartening—the degree to which the resulting trial, known as the Pig Trial, gained worldwide attention. Activist Louise Jorgensen organized courthouse vigils for every pre-trial and trial. I did not feel like I was fighting this alone. Chris Foott, organizer of pig vigils in Manchester, UK, came to show his support along with his wife, daughter, and mother. He said to me, "I came to support you at the Pig Trial and to support the whole movement and whom we were all standing for—the pigs." I was grateful for all the love and solidarity shown by the global network of supportive animal rights activists and organizations, including PETA and Animal Justice.

James Silver and Gary Grill, my two vegan defense lawyers, worked hard *pro bono* to put the pig vigils on the map. They devised a strategy to turn the tables and put animal agriculture on trial. They called on a host of expert witnesses to address all the reasons for the vigils, including the suffering of the pigs, animal rights, animal sentience and personhood, and environmental and public health concerns surrounding animal agriculture. Our team presented two main arguments: first, that compassion is not a crime and giving water to thirsty animals is part of the universal Golden Rule that transcends time and culture; second, that pigs are not "stuff" or property but individuals deserving of personhood status.

In August 2016, Silver and Grill invited an expert witness, veterinarian Dr. Armaiti May, to testify. Dr. May said that pigs in transport trucks suffer from extreme heat stress and dehydration during the summer, which she said was likely the case with the particular pigs concerned by the trial. She observed that some of the pigs were foaming at the mouth and in "severe distress," appearing to breathe at as rapid a rate as 180 breaths per minute.[16]

Compassion Is Not a Crime

My case resonated with the public. People understood that offering water to pigs on a sweltering day constitutes an extension of the Golden Rule: "Do unto others as you would have them do unto you." Social media users sprang into action with their support. By the end of 2016, a care2 petition called "Compassion Isn't a Crime: Giving Thirsty Pigs Water Isn't a Crime" had been signed by 300,000 people, sparking the social media hashtag #CompassionIsNotACrime.

---

16 Expert Witness Report of Armaiti May, Regina v. Anita Krajnc, No. 1211998163042 (Can.); Expert Witness Report of David Jenkins, Regina v. Anita Krajnc, No. 1211998163042 (Can.); Expert Witness Report of Lori Marino, Regina v. Anita Krajnc, No. 1211998163042 (Can.); Expert Witness Report of Anthony Weis, Regina v. Anita Krajnc, No. 1211998163042 (Can.).

Worldwide support included protests at Canadian embassies in Argentina and Portugal, and vigils held in solidarity, including ones organized by Save chapters in Ottawa and St. John's in Canada; in Arkansas and North Carolina in the United States; in Manchester, Essex, Bristol, Guildford, Spalding, and West Midlands in the United Kingdom; in Rome, Italy; and as far afield as Melbourne, Australia. Incredible artwork poured in from a score of artists, including Alba Paris's emotional drawing of a tearful pig approaching a woman's hand, held out to offer water. American activist Ryan Philips published a graphic novel titled "Anita's Story: Compassion Is Not a Crime." In Burlington, we held a special vigil in which activists wore large stickers that read I AM ANITA on their backs while giving water to pigs.

Mainstream media coverage included videos and photos of us giving water to the suffering pigs, thus allowing the public to see the animal victims in the same way activists see them on a regular basis. A handful of reporters joined our vigils, bearing witness firsthand alongside activists. Karin Wells, a senior reporter for the CBC (Canada's Broadcasting Corporation), did a half-hour documentary for the radio show *The Sunday Edition*. While capturing the street sounds of a vigil in front of Fearmans slaughterhouse, Wells interviewed vigil organizer Lori Croonen about bearing witness.

Karin Wells's narration goes: "Suddenly, heads turn towards the highway. 'Two trucks coming in. Two trucks with pigs.' [You hear activists speaking and the truck braking.] They come down the hill, catch the red light, and stop. Two of the people on the traffic island reach into the truck through the open holes and stroke the pigs.

'Hi angels.' The pigs move around, pushing their pink snouts towards the hands. 'Hi sweetheart. Hi sweetheart. Yeah. I know. I know. You shouldn't be in here. You shouldn't be in there, sweeties. We're fighting for you, okay? We're fighting for you. Yes, you're a good piggie,' [says activist Lori Croonen]. Tears pour down the woman's face.

It makes you sad [says Karin Wells to Lori]. 'It does all the time. But I haven't been able to face it until recently. Now, I'm like, you just got to do it, right?' [responds Lori].

The driver watches in his rear-view mirror. The light changes, and he turns left and drives the pigs into the slaughterhouse."[17]

The media coverage moved quickly from the local level to the national, then international level as early as in the pre-trial stage. There were media scums in front of the Milton courthouse. A *Canadian Press* story by Liam Casey jutted the case onto the national stage when it appeared in major Canadian newspapers. Casey set up the conflict as one between factory farmer Van Boekel, who referred to the pigs as property ("just leave my stuff alone"), and the defense with our raised moral arguments ("I think it's an outrageous charge that goes against my deepest philosophical beliefs in terms of what all our obligations are, and to me, the most important thing in life is to be of service to others and to someone or some animal who is suffering").[18]

The story was covered in all the major British newspapers, then spread to Hong Kong, the *Russia Today*, Norwegian and Polish television, and many other international venues. We had some celebrity support. Maggie Q (star of *Mission Impossible III*, *Live Free or Die Hard*, and TV show *Designated Survivor*; co-producer of the classic animal rights film *Earthlings*) held an early-morning media conference at Dark Horse Espresso on Spadina Avenue in Toronto on November 1, 2016, hours before the trial resumed: "When I first heard Anita's story, I thought it must be a joke. There's no way that someone could actually get arrested—let alone be facing jail time—simply for showing a little compassion to a suffering animal in the final agonizing moments of an unjust life."[19] Ingrid Newkirk, founder of PETA, attended the court hearing and told Catherine McDonald of Global News television: "We think of kindness as a virtue, not as something to be punished. This is rather unique because no one can really understand

17 The Sunday Edition, "Is It a Crime to Give a Pig Water on a Hot Day?" CBC radio broadcast, March 20, 2016.

18 Liam Casey, "Unapologetic Ontario Woman to Appear in Court for Feeding Pigs Headed for Slaughter," *Toronto Star*, November 4, 2015.

19 Zachary Toliver, "Actor Maggie Q Voices Her Support as Trial Resumes for Activist Who Helped Suffering Pigs," *PETA*, November 1, 2016; Justin Skinner, "Anita Krajnc, Maggie Q Speak out about the Treatment of Animals," *Inside Toronto*, November 1, 2016.

how you can bring a charge like this. Jesus gave water to a slave at the well. Do you expect him to be locked up for that? It's just absurd."[20]

When I took the stand, I was briefly cross-examined by Crown Attorney Harutyun Apel. He asked me how I would feel about a stranger offering water to my dog, assuming a dog could be viewed as property. I pointed out that my dog, Mr. Bean, is very assertive, and that if a stranger offered a thirsty Mr. Bean water, we (he and I) would gladly accept.

In *A Calendar of Wisdom,* Tolstoy states that the Golden Rule applies to all living beings: "We should take pity on animals in the same way as we do on each other. And we all know this, if we do not deaden the voice of our conscience inside us."

The sheer irony of compassion being treated as a crime creates resistance, helping to further educate and mobilize people to stand up to injustice. A historical example: In 1892, the Russian government attempted to stop Tolstoy and his family from engaging in famine relief by making it illegal to set up soup kitchens for the hungry. Tolstoy's response to this absurdity was, "People cannot be prohibited from eating." The police officer charged with carrying out this law explained: "Put yourself in the position of a man who is under orders from his superiors. What would you have me do, Your Excellency?" Tolstoy offered this rebuke: "It's very simple: don't work where you can be made to act against your conscience."[21]

Pigs Are Persons, Not Property

Our most important aim was to try to get people to see the pigs in the transport trucks from a perspective they hadn't considered before. Indeed, it can be difficult for a non-vegan to see nonhuman animals as persons. But we could at least get them to see a farmed animal as "somebody, not something," not livestock or property but a sentient individual. It wasn't just me who was on trial; these beliefs were as well.

Dr. Lori Marino, a neuroscientist internationally known for her work on animal behavior, brain function, intelligence, and sentience, also gave

20 Catherine McDonald, Global News Television Broadcast, October 3, 2016.

21 Ilya Tolstoy, *Tolstoy, My Father: Reminiscences*, translated from the Russian by Ann Dunigan. Chicago: Cowles Book Co, 1971, p. 229.

testimony in the Pig Trial. In Dr. Marino's submission, she stated: "Perhaps more than any other psychological characteristic . . . self-awareness is most intimately tied in with the ability to suffer. Self-awareness is the property of having a sense of ["I"] and the ability to think about oneself and one's circumstances. . . . [Pigs] have been observed making repetitive movements while appearing to watch themselves in front of a mirror. These movements, called ["contingency checking,"] are precursors to mirror self-recognition. . . . Second, pigs can use mirrors to find food hidden from direct sight but only observable in the mirror. . . . Pigs, like chimpanzees, understand that a joystick they control moves an on-screen cursor. . . . Pigs have the cognitive ability to play a video game."

Dr. Marino continued: "Taken together with the other capabilities described above, these findings are compelling evidence that pigs are self-aware beings who not only have a sense of self but understand their own body in relation to the environment and can even use that understanding to achieve goals. This kind of animal, with the capacity to not only feel pain, distress, and hardship but also to have the intelligence and wherewithal to know that [their] life is not going well, is particularly vulnerable to experiencing suffering during inhumane conditions like those depicted in the video of dehydrated, overheated, overcrowded, and frightened pigs during transport to slaughter."

In his examination of Dr. Marino, defense lawyer Gary Grill asked, "Are pigs persons?" Marino's reply was "yes." She then testified that pigs are autonomous beings; they are highly self-aware; they are sentient; and they possess complex emotions.

When nonhuman animals are legally considered "things," their basic, inalienable rights and fundamental interests—their pains, their lives, and their freedoms—are invisible to civil law. But since they suffer the same as we human animals do, they deserve equal consideration of their interests and needs.

When I took the stand in October 2016, a key part of my testimony discussed who pigs are. For that, there was no better example to present than the pigs at The Pig Preserve. In court, we showed a video I had taken of actress Lauren Maddox and Richard Hoyle discussing the habits and language of the 120 pigs roaming freely on the Preserve's 100 acres of grassy hills and woodlands. The video focused on a few pigs, including Blue Eyed Wilbur and Rosebud. Then, for contrast, we showed a gruesome video depicting how billions of pigs are treated each year in industrial animal farming. We used an Aussie Farms (an Australian nonprofit) video that captured pigs being brutally electrically prodded into the carbon dioxide gas chamber to be slaughtered. I made the case for the universal duty to bear witness—to not look away when an animal is suffering but to come as close as possible and try to help, to paraphrase Leo Tolstoy.

My final court hearing took place in March 2017, almost two years after the incident. On the day of the verdict, so many activists came to show support, some from very far away, like journalist Jane Velez-Mitchell and *Earthlings* director Shaun Monson from LA. Child actress Mckenna Grace and her mom joined us, but not before bearing witness in front of Fearmans slaughterhouse, where she said gentle "I love you's" to the pigs and wondered why the police weren't helping to stop the injustice. She gave out her little homemade heart- and pig-shaped cutouts with the words Compassion Is Not a Crime. I still have mine on my dresser.

The activists in the courtroom applauded when Judge David Harris dismissed my charge of criminal mischief (interference with property). I was innocent, he argued, because I hadn't stopped the truck or prevented those pigs from being slaughtered. It was a victory of sorts that he recognized that compassion is not a crime. Regarding our second key objective, the judge's ruling was a massive disappointment. Judge Harris missed an opportunity to move the law forward on the rights of nonhuman animals as persons. Instead, he maintained that pigs were property under the law.

Everyone knows animals are sentient. Is our legal system so far behind that a sentient being, such as a pig, is treated no differently from a toaster? This injustice is only amplified by the fact that non-living entities like corporations have legal personhood, or standing, in the courts.

4

Animal Save Movement Develops Roots

When my confrontation with the truck driver took place in the summer of 2015, we had only around thirty-five Save chapters in Western countries. After the Pig Trial generated a wave of international attention and, subsequently, financial support, Animal Save Movement began to focus on a growth strategy. Organizing drives are necessary to speed up the movement-building process. We did this by holding more animal vigils, by sending organizers to table at VegFests, and by organizing tours to set up new chapters. There are millions of slaughterhouses worldwide; our aim was for the animals being sent to slaughter to be recognized as individuals and no longer go unnoticed. We hoped to develop a grassroots movement, with the goal of having vigils at every slaughterhouse. We felt that people would engage in a Gandhian, "noncooperation with evil" movement if only they knew about it.

New groups around the world formed organically, bearing witness to more and more animals, including pigs, cows, chickens, turkeys, goats, sheep, rabbits, horses, and fishes. But this was a slow process. Often, visitors from another town or country would attend one of our pig vigils and get inspired to form their own group after the intense experience of watching truck after truck filled with pigs arriving at a slaughterhouse.

Many of the groups that formed five to ten years ago established deep roots and continue to hold vigils and engage in other types of activism, exercising innovation in order to grow. There are now pockets of stable Save chapters in Canada, the United States, the United Kingdom, the Netherlands, Italy, Portugal, Spain, Mexico, Peru, Argentina, Uruguay, Turkey, Israel, Kenya, India, Australia, and New Zealand, among others.

The Early Save Chapters

Since 2016, Manchester Pig Save has been holding vigils in the nearby town of Ashton-under-Lyne in Greater Manchester—at a slaughterhouse called Tulip. Among its organizers, Andrew Garner, an activist in his mid-twenties, had been keen to start a Save chapter in Manchester since the end of 2015, around Christmas time. Andrew had been active through Earthlings Experience, a form of activism in which activists stand in downtown areas and hold video screens showing slaughterhouse footage to try to convince onlookers to change their diets. "We were brainstorming on what more we could do. There were many single-issue groups covering fur and animal experimentation, but we wanted to redirect focus onto vegan education. There was something missing." That was when Andrew and his co-organizers discovered Toronto Pig Save. The choice of slaughterhouse seemed easy: Andrew knew of the Tulip slaughterhouse because his aunt and cousin worked there.

The organizers visited the slaughterhouse, scouting areas to hold vigils and speaking to security. There is a wide turning lane for trucks to enter, which proved to be a good location. Nearby, there is a gate that leads to a country footpath. In the evening, they walked along the footpath, which runs alongside a canal behind the slaughterhouse. They could hear screaming and realized it was coming from outside holding pens and "kill floors." As Andrew recalls: "It was a life-changing experience just to hear [the animals]. We needed to start a Save group there. People need to know what's going on on their doorstep." Manchester Pig Save's first event was organized on a Saturday in February 2016. There was a big turnout of around fifty people. The *Manchester Evening News* covered the vigil, but there were no trucks. The slaughterhouse was closed on Saturdays. Nonetheless, people came from as far away as Liverpool, London, Newcastle, and Essex, having driven many hours to get there.

Andrew says: "At first, the slaughterhouse management didn't know what to do with us. They expected us to be aggressive, but we were doing peaceful, love-based activism." It took a few vigils before they secured an arrangement for the trucks to stop so activists could bear witness. At one point, when one activist decided to climb a truck and stand on top of it with outstretched legs and arms, the management didn't want the vigils to continue. Andrew took it upon himself to persist and work with the security guards and police to ensure the trucks stopped again. Eventually, the police made this happen, assessing it to be the safer option.

Andrew also helped with video documentation of the vigils. His first video captured the connection he made with a particular pig, who calmly looked at him and moved towards him, despite the bustle in the truck. With the knowledge that "this [was] another soul who [would] go to a gas chamber," Andrew burst into tears when he got home. Upon posting the video, he remembers that "quite a lot of people saw [it] and read [his] caption," including "a whole family of four [that] went vegan from that interaction."

Andrew's aim was to get people to not only go vegan but also become organizers. Tyler, a graphic designer from Essex, had been attending the Manchester vigils, though that meant having to take an overnight coach each time to get there. Andrew convinced him to start vigils in his hometown. An earlier group had held protests there following an exposé in which slaughterhouse workers were shown abusing pigs on the kill floor in horrific ways, including putting out cigarettes on their faces. Andrew said he'd come along and help Tyler talk to security, management, and police. Essex Pig Save, the second Save group in the UK, was thus born. Andrew also had a hand in starting Liverpool Pig Save, along with his partner.

Chris Foott was in attendance at the first Manchester Pig Save vigil in 2016 and continued to attend all the vigils, including special, all-day events. Sometimes, he would head straight to a vigil after his night shift had ended. A firefighter for well over a decade, Chris is conditioned to see death as normal. Yet, he says: "When you see these innocent victims and look them in the eye, it creates a unique set of emotions. It's different from my job. You're face to face with hundreds of individuals. You know whatever you say to them, whatever you do, these hundreds of individuals are moments before death. . . . Seeing hundreds of [nonhuman] people before their deaths is pretty emotional."

The pig vigils were his first taste of vegan activism and moved him unlike any other experience had. "You were on the frontline, you were face to face with pigs, you were there comforting them," Chris says. "Lots of people who are vegan are quiet about it. That isn't enough. It's about encouraging others to get involved. Some activists are online and speak up from home. For us, it was being on the frontline. The moments before the animals go to slaughter are the most crucial ones. It's our last opportunity to see the animals and to make them visible to others. For most animals, it's the only compassion they'd see from humans. The only time they'd be exposed to the outside world. For most of us humans, it's the only opportunity we have to come face to face with them."

The first vigil was held before dawn on a cold and wet winter day. Despite that, lots of people came and gathered on the sidewalk close to the entrance in the pitch-black darkness. At first, the lorry drivers and security guards were resistant, refusing to stop and let activists bear witness. Activists held up signs that read 3 MINS IS ALL WE WANT. Some trucks began to stop, and people would rush over to get a glimpse of the poor pigs. Several months later, the group and the slaughterhouse would successfully come to an agreement that allowed the trucks to safely stop.

Manchester Pig Save devised strategies to increase attendance at vigils and invite locals to join. They set up all-day vigils on weekdays, which would draw over a hundred attendees. Social media influencers like Earthling Ed, James Aspey, and Bite Size Vegan were invited, thus bringing in lots of new people who, despite not having been involved in activism before, would suddenly decide to attend their first vigil. In a nearby field, to the side of the slaughterhouse, a gazebo was set up for activists to have vegan barbecues and picnics.

The public footpath between the slaughterhouse and the canal that flowed along its side was used by the locals to walk their dogs. These people might have heard things but never given it much thought. However, as Chris attests: "Activists made the local community of dog walkers more aware of what's happening. If you get close enough, you can hear pigs scream. It was like listening to hell. We asked people to come closer, not even to see the pigs but just to listen to their cries. Locally, it raised awareness. It encouraged more people to go vegan. It encouraged families to go to vigils. My whole family attended the vigils." Chris remembers attending a Toronto Cow Save vigil with his parents when they visited Toronto: "It just opened up all the emotions. Dad was turned almost vegan when we attended a cow vigil there. After the vigil, we went to a café nearby, a Jamaican place. . . . He got involved in conversation. After that, he stopped eating meat. Toronto Cow Save made my dad go vegetarian."

Manchester soon became an activist hub, and people who traveled or regularly attended vigils there began holding vigils at other slaughterhouses. Chris helped fellow organizer Kate set up Bolton Chicken Save half an hour outside of Manchester, at a slaughterhouse that was open only at night and killed about 10,000 chickens once a week. These chicken vigils were exceptional. Activists were able to get so close to the scene that they could see the chickens being hung by their feet on the moving conveyor belt. Initially, the workers tried to block vigil attendees by putting crates in the way and even spraying water at them. When the police were called, they were supportive of the activists. Eventually, the relationship changed. Activists were able to get friendly with the son of the slaughterhouse owner and build trust with the staff. Chris recounts: "We'd be able to rescue some chickens. Sometimes, they gave us chickens. Sometimes, they let activists hold the chickens. If we do more compassionate and love-based activism, it is better for everyone." Jennifer O'Toole corroborates his story: "I was attending a

Bolton Chicken Save vigil, where the crates holding the chickens are right on the street. The workers allowed activists to open the crates and show love to these animals while others, just meters away, were being hung up on the conveyor belt to be slaughtered. While we were bearing witness, a chicken leapt into my arms from the crate. I held her in my arms for a long time and didn't want to let her go. The workers eventually agreed to surrender her to me and I was able to rescue her."

James O'Toole, one of our communications directors and a former stockbroker, started his activism with Toronto Cow Save. It was the first activism of any kind that he had ever done. He recollects: "It's not something I felt comfortable with at all. One of the things I liked about Save is it wasn't very confrontational; it returned love for hate. That approach sat well with my personality. If you are arguing with people, they instantly close their minds. With love-based activism, and if you're calm and you communicate with people, it's the most effective form of activism." James was vegan for three years before he became an activist. "Being an activist is something I never thought I'd do."

He first bore witness by accident in 2013 when he was cycling by a Toronto Pig Save vigil. He lived nearby and was cycling downtown. James saw the pig vigil from across the street. He could see the pigs in the trucks and the activists giving water to and comforting them. "Because I saw a truck, I thought differently of the activists. Normally, if I just saw people protesting, I would think they were crazy people. Historically, that is what I thought of activists. However, my perception was totally different because the animals were there. It stayed with me—this nagging feeling of wanting to go see the animals. Even though I hadn't been able to properly bear witness (I was across the street), it was enough to motivate me, months later, to go to the cow vigil." That was in 2014. "The very first time I came face to face with the cows, it transformed me. These cows demonstrated all the emotions that humans would in the same situation: defeat, sadness, fear. . . . Nothing was more important to me than stopping animals from being exploited." He succumbed to the power of bearing witness. "Previously, I'd seen the PETA video 'Meet your Meat,' but it hadn't inspired me to be an activist. I saw that and thought, *I'm a vegan already*. I don't know how you could not be greatly affected by the animal victims if they were in front of you. It's a transformative experience."

James started helping us with social media, including building the Toronto Animal Save Instagram accounts into a formal presence by posting daily content from the animals' perspective. Many of our social media posts have comments from people saying they are going to go vegan because of whom they see in a photo or video; however, social media does have its limitations. "It's not the same as being there, face to face. Nothing can compare to that. Nothing motivates people to get active for animals more than bearing witness."

James expands on the community-building power of the vigils: "As I'm an introvert, community wasn't something that was important to me at the time. However, the organizers were very welcoming. After the cow vigil, they would invite everyone to Second Cup Café. I felt that a sense of community was very helpful. That provided comfort after witnessing all the atrocities. Any way we could build up the community, our strength in numbers—it's an organic thing, that sense of togetherness. There are a lot of vegans who are not activists . . . so if they came, they'd experience a sense of community, and obviously, it would help the animals."

While attending South Florida VegFest to help set up new chapters, James and his wife Jennifer had the opportunity to go inside a slaughterhouse: Mary's Ranch in Hialeah. What he concluded, however, is perhaps counterintuitive: "Seeing animals in trucks in front of slaughterhouses—seeing their fear and their emotions—is in some ways more powerful than being inside a slaughterhouse. Slaughterhouses are more horrific, but bearing witness before [the animals] go inside the facility can be the worst experience. When I saw a pig getting slaughtered in front of me, it was horrible. But at least their lifelong suffering was over. But knowing that someone is waiting in line, that weighs on your mind more. . . . I just spent five minutes with a cow, and I know what's to come for that cow. It's all ahead. For some reason, that's hard to process. . . . You want to stop it and save them. It's a different experience."

Azul Cardoza became vegan in 2016 and started activism in the same year. She dove into animal rights workshops as well as street activism such as Earthlings Experience screenings, Cubes of Truth, and Direct Action Everywhere (DxE) disruptions. "It was more difficult because of the place where I lived," Azul shares. At the time, Azul was living in Medellín, Colombia. After receiving repeated encouragement, Azul began opening Save chapters all throughout Colombia in 2018. She explains: "In Latin America, all the slaughterhouses I visited are located in areas of extreme poverty. When we are there, we are connected with the animals. The residents living in the vicinity also want it to end."

Azul organized vigils at the biggest slaughterhouse for pigs in Medellín, called Porcicarnes (translated as "pork meat"). Activists spoke to security at the gate and said that they were there just to take photos. At the first vigil, they bore witness to nine huge trucks, each with two floors "full of these beings who were so sad." "Activists held hands, surrounding a truck. I was already an activist. For me, it was a different experience. I felt a lot of sadness and connected with the animals. It was the first time I connected with an animal. It's a unique experience." Surrounding a transport truck is a very creative nonviolent direct action at vigils. Similarly, on January 28, 2018, Buenos Aires Animal Save activists performed a symbolic hug around a transport truck full of cows waiting to be unloaded at the slaughterhouse. The cows were initially nervous but were eventually calmed by the contagious energy of the hug. The livestream of the action went viral, with over a million views on Facebook and coverage by several mass media outlets.

Like James, Azul says: "Vigils have a power that no other form of activism does—this connection with the animals, this conviction it creates. . . . So I think it's different. When people come to vigils, it's really important because society uses propaganda. We never see animals suffer. This is the only connection we can make. For activists, it's different. You can channel all this anguish into more strength. That's why it's important. It's a revolutionary act of love and resistance." Unfortunately, over time, security kept asking activists to move further away from the trucks and made bearing witness more difficult.

Ana Carranza is a country liaison for Mexico, having organized in Guadalajara and now organizing in her hometown of Hermosillo. Having grown up on her family's ranch, she remembers seeing a truck being loaded with cows one night, which left a strong impression. Before that night, she had not realized that the cows would be killed. She was shocked. Decades later, Ana, a vegan activist, bears witness to cows in slaughterhouse trucks and, like many others, finds it to be a uniquely persuasive form of activism. "It is very powerful to be directly with the victims. In other forms of activism, we just talk about how systems work. But for vigils, you are meant to be with the animals. You are seeing the victims directly. You are taking photographs. The industry wants no one to see the eyes of the animals. They don't want anyone to know that they exist." Ana has had success talking to the truck drivers, getting some of them to stop and allow the activists to be with the animals. "I don't know why, but I think it's because we treat them with respect. We always ask the drivers if we can be with the animals. Just once in all these years, a driver didn't want us to be there. Most others give us a lot of time, sometimes fifteen minutes. We take photos and videos."

Valezca Munsuri, also a country liaison and an organizer for Mexico City Animal Save, reveals that in Latin America, there is a pattern of slaughterhouses being situated in communities that face extreme disadvantages. "There's a prevalence of poverty, disease, and violence, mostly violence against women. In Mexico, the laws that 'protect' animals exist but don't get enforced. We see animals being hit, who are bleeding and very thirsty after spending hours under the sun. I often see some individuals in the trucks who are already dead." The first vigil Valezca ever attended was in Mexico City in 2017, put on by Save Tour organizers Paola and Fabian, who traveled from Lima. Valezca was familiar with the vigils in Toronto, so when she learned about the event, she knew she had to be there. And when Paola asked, "Hey, who wants to open a chapter?" Valezca immediately responded, "I want to join and organize."

Together with Paola, she investigated several slaughterhouses, ultimately choosing a pig and cow slaughterhouse where vigils continue to be held today. As Valezca points out: "The logistics are good. The slaughterhouse isn't in a residential area, but people have houses nearby. It is safer to stop the trucks because there is only one way to get in and out of the slaughterhouse, and it's not on a freeway."

Valezca recalls her first experience connecting with the animals in the trucks: "The first time we stopped a truck, it was a pig truck. We were nervous. The driver was friendly and stopped. There were twenty activists there. We approached the pigs, who were screaming and crying. They were really thirsty because it was a very hot day. When they realized we were giving water, pigs from a higher level of the truck stuck out their snouts. My boyfriend pulled me up on his shoulders so I could reach these pigs and give them water. The driver gave us five minutes, then we had to step aside. We were overwhelmed with sadness and in shock. Then, more and more trucks started arriving. We had to focus. We had to accomplish objectives and remember why we were there: to ensure the security of activists, give water, take pictures, and livestream to social media. I still shake after I see animals on a truck. After all these years, that has not changed."

Valezca and Azul are also directors of recruitment for the global Animal Save Movement. Once a month, they host welcome parties for new organizers to strengthen the sense of community within the Animal Save family. She says it's a constant effort to get more people to attend vigils. But her group has had an impact. She is particularly proud of the group's effort to hold more frequent and regular vigils, which got the attention of the local media. "They reported what we witnessed. They shared images of the animals bleeding and suffering in the trucks and images of local activists trying to help. The owner of the slaughterhouse was really mad. He went outside and waved the newspaper in front of us. 'You are messing with my work because farmers don't want to bring cows and pigs here to slaughter anymore. We are losing money.' We had affected their work. There are not as many trucks anymore. We used to see ten or more trucks in a space of four or five hours, but now we see only three or four trucks, sometimes just one or none at all." One of the organizers, Avril, who lives nearby, says: "During the week, not so many animals arrive. The COVID-19 situation also affected the slaughterhouses because they are really low on both staff and animals now."

There are several safety marshals at the vigils, including one whose role it is to talk to truck drivers. When Valezca has this job, she takes the opportunity to ask drivers why they are doing this type of work and to do vegan outreach. Once, at a vigil, she saw that a driver she had talked to at a previous vigil had bought the plant-based food she had recommended. "Now, they are not eating animals anymore. We give truck drivers and slaughterhouse staff plant-based meals and do outreach between trucks. They tell us how the animals fight for their lives. They don't want to die. 'We understand why you are here,' they say. Mostly, we see positive reactions."

"The slaughterhouse is in a very poor and violent district. At the end of the street, there are houses and a restaurant. Families live nearby and children join us. We talk to them. We gave a young girl—around eight years old—some chalk. She drew animals. She was always asking us about animals. She said she hears the pigs screaming at night. They kill animals twenty-four hours a day at this slaughterhouse. The screams don't let her sleep and she gets really scared. We shared vegan food with her." Thinking of the poor girl, I asked Valezca how she and her fellow activists can create peace in a place where there is so much violence, where there are so many slaughterhouses. She told me they are planning on building a community garden and giving away seeds so that the locals can grow vegetables and fruits. Though restaurant owners sometimes get mad at activists for stopping the trucks in front of their businesses, Valezca's group has undeniably made a difference. "The community is affected by the slaughterhouses. They are so sick because of the air and water pollution. Locally, we have had a positive impact. Now, they are friendlier. They send us water and juices."

At the end of each year, in December, Mexico City Animal Save activists take to the city center; they close a main avenue to hold demonstrations and performances and show the public what they have documented. The campaign is called "Mexico Without The Slaughterhouses." Valezca and her team screen footage from their investigations on a big monument, garnering attention from mass media. "We know it's not enough to just hold the vigils, so we are doing actions in the city because we know the majority of people who live there are very far from this slaughterhouse. We take the slaughterhouse to the center of the city."

North Carolina Farmed Animal Save formed in 2013 to hold vigils at the Smithfield slaughterhouse in Tar Heel. The state is known for its intensive farming of pigs. The massive slaughterhouse employs around 6,000 workers and kills up to 33,000 pigs each day using three Butina gas chambers running side by side. At vigils, you can see truck after truck full of pigs coming into the slaughterhouse all day, lined up almost bumper to bumper. North Carolina has the second-largest concentration of pig factories in the US, after only Iowa. There are far more pigs (10.1 million) than humans (9.4 million) in North Carolina. Organizer Roxanne Kirtright explains: "It really is the worst place for a factory-farmed animal. I am on the southeastern tip of the state, and according to *The Final Nail* [a website that lists slaughterhouses in the US by state], there are fifteen slaughterhouses in my area code, including the largest pig slaughterhouse and largest turkey slaughterhouse in the world."

Steven Wise, founder of the Nonhuman Rights Project, which advocates the granting of legal personhood and the corresponding rights to various animal species, recalls his experience visiting a Tar Heel slaughterhouse: "Well, it's something I haven't forgotten. It makes concrete the things I might have read about, even the films or pictures I might have seen. You can actually see and feel and smell—and sometimes touch—the animals as they are coming into the slaughterhouse. I was able to do that at the small slaughterhouse. I was even able to go up to a calf and put my arms around her and hug her on the way to her death. That was in the 1980s—no one knew who I was and they didn't care that I was there. In the twenty-first century, they knew who I was and they cared that I was there, so that's why I wasn't able to do that. I was able to find out what was going on in the slaughterhouse only by putting an ad in the local newspaper in Tar Heel, saying that I would pay slaughterhouse workers to come and see me at a certain restaurant at a certain time. I wanted to interview them. People came, and so I paid them to sit there and speak to me about what they did and what was going on in the slaughterhouse in a general way."

Animal Vigils Are for Everyone

When you consider starting a Save group, ask yourself: "*Where are the slaughterhouses in my community?*" They can be found in the countryside, in downtown cores, and in city suburbs. For example, Fearmans Pork Inc. is located at a main intersection in Burlington, Ontario, making weekly vigils at this facility highly visible. The experience at each slaughterhouse is different. At some vigil locations, such as the Tulip slaughterhouse in Manchester, the sounds made by the pigs inside the gas chamber, crying out for help, are audible. Similarly, at Conestoga Meat Packers in Breslau, Ontario, the pig vigils are held on a public road near the kill floor, and activists can hear the horrible screams of the pigs being prodded into the gas chamber. Tricia Brubacher, a local activist, recalls hearing this sound of torture at her first vigil and going vegan on the spot.

Boxtel Pig Save in the Netherlands established a rather unusual agreement with the slaughterhouse where it holds vigils, whereby activists are issued passes, which they have to show to enter and stand in a designated area on the slaughterhouse property. The slaughterhouse provides bottles and a big blue bin of water for activists to refill their bottles. Organizer Lea Goodett says: "When I started doing vigils, it was stereotypical activists with nose rings. They shared their stories with other activists. It was a bubble. Over the last few years, we started seeing more mainstream people. They are not activists. They are just curious." Lea says what helped was inviting social media influencers to attend vigils, who would then share images of the pigs in the trucks with their followings. "It created a major shift. People who followed influencers started attending vigils. That's what we want . . . to reach outside of the activist bubble. The people who are curious but who are not yet doing activism. At every vigil, half of the attendees are familiar activists and the rest are new people!"

Lea puts a lot of effort into reaching new people through the group's newsletter, but still, at the beginning, very few people wanted to join the vigils. "Once I got the first social media influencer, that changed." The first influencer to attend the Boxtel vigils was @lisagoesvegan. In December 2020, after attending a vigil, she posted images of the animals on her Instagram, next to super pretty food posts. Lea reports: "These vigil posts got so much attention. Followers appreciate the way their influencers show their emotions to the public. When they film themselves at the vigils, it becomes real for people. People care about the influencers crying and how sad it is to look at the pigs. It matters that it's their footage and images from the vigils. They are sharing 'softer' images of the suffering animals—in the sense that there is no blood like on the kill floor. But it is so sad. A lot of knowledge is not out there. People say, 'Oh my God, I didn't know that this slaughterhouse kills 22,000 pigs a day or 300,000 chickens a day.'" Despite fears of losing followers in posting such sad content, influencers find the opposite happened. Followers engaged more with vigil posts on social media by a factor of several times compared to other posts.

The origins of Las Vegas Animal Save lie in an LA Animal Save pig vigil in 2017 at Farmer John slaughterhouse. It was the first time Camille Savage had seen a slaughter truck up close, with animals on it. "It was a really creepy atmosphere at night, with the wall murals." There are painted happy pigs, frolicking and flying, for blocks around the perimeter of the slaughterhouse. Camille recalls: "Farmer John is a huge compound. It's like you're alone in a crowded room. In the moment, it feels like it's just you and the animals. I felt immense sadness." When she returned home, she and fellow activists discovered two slaughterhouses in Las Vegas. They weren't easy to find, Camille explains. "I had no idea there were slaughterhouses. Unless you go on delivery day, you wouldn't know it's a slaughterhouse." The day after Christmas in 2017, Camille started social media pages for her new group; and on January 6, 2018, she organized the first vigil at Highland slaughterhouse. Las Vegas Animal Save now holds two vigils a month.

Camille describes the experience of bearing witness: "Never forget those we might not see because they are out of our line of sight—they are there and they need us. What's happening needs to be exposed." She further notes: "Frustration is the root emotion, which in turn leads to sadness and anger. It's upsetting because we want to take the chickens out of the truck and give them the life they deserve. But we know we are bound by these unjust laws. We have to be strategic and use the opportunity to expose the injustices. I'll be damned if we don't do our best to tell their stories, to rescue them when there's an opportunity, and to talk to slaughterhouse workers that we will continue to be here because they do matter. Vigils never get easier." And Camille is right: the frustration is monumental.

Building Inclusive Communities

Let's returning to Manchester for a moment. These were the first Save Movement vigils in the United Kingdom; even so, they continue to grow and be held consistently. Josephine Robinson, who founded Bristol Animal Save, praises the "wonderful community-building skills" of Arun and Silvia, who took over as the lead organizers in Manchester, who "make sandwiches and put up prayer flags all around the entrance of the slaughterhouse." They introduced a social aspect to the vigils. In 2021, the band Mobius Loop came and played a fitting song, "May All Beings Be Free," outside the gates of Tulip slaughterhouse. The Hare Krishnas have held a concert. Because the Krishnas are vegetarian and not vegan, Arun and Silvia asked that they do the blessing with plant-based milk instead of dairy milk, which they happily accommodated.

Arun and Silvia had been attending the Manchester vigils for a couple of years but didn't get a sense of community. Their first task as lead organizers was to rectify that. "When we first attended, we felt very isolated. It's not that people weren't friendly, it was a bit cliquey. We found it hard to integrate." They had ideas to expand the chapter. "Before we can bear witness, the focus has to be on people. We've got to build a community people want to be a part of. We have to welcome people, organize social events, and create friendships." Also, Arun says that when there's conflict, he tries to take a neutral stance to be supportive of everybody.

Silvia took on the job of reaching out to all new attendees. "We spent most of our energy following up with activists and making sure they are coming back. We welcome them to our vegan family." In the year after they started organizing, they saw a number of new people attending the vigils. Silvia would spend time contacting newcomers on Messenger and checking in to see how they felt after a vigil. Within half a year, a new solid core team developed. Silvia began organizing Zoom meetings for new attendees to make them feel included.

Arun also points to the need to spot and value every activist's individual talents so as to resist falling into a hierarchical organizational structure. "We can grow Manchester Pig Save into a mass movement. Give people a chance to shine. We do that to all our activists. We get to know them. We see what they can bring and what they can do beyond holding placards. . . . We are saying, 'Here's the canvas.'"

"If you just bear witness, you will get burnt out or get bored." With this understanding, Arun spices things up (literally!) by making the security guard curry. That same security guard, in turn, bought everyone drinks when they went out. "Our aim is to build a mass movement. Lots of new people are joining the vigils and many are becoming vegan activists. We are growing bigger and bigger. Activists are bringing their moms and local people are chatting with us."

Another key takeaway for Arun is that organizing has to be inclusive to be successful. The first step is finding common ground with the wider community. The chapter emphasizes being a compassionate, inclusive social justice group that cares not just about animals.

Many vigil organizers have stumbled into the pitfall of attracting only activists within the vegan bubble. It is not immediately obvious how to make activism inclusive. Jane Velez-Mitchell, former CNN anchor and founder of UnChained TV, has consistently livestreamed the Los Angeles pig vigils. Even so, over several years, she has been able to get only one or two of her nonvegan friends to go to a vigil. One friend, who is a self-help guru, says he can't go to vigils because it'll make him sick. Another friend, incredibly, told her, "I don't want to see a snuff film." "People have a real aversion to witnessing this. It puts the abuse they are a part of in their face." That's why videos are so important, says Jane. By viralizing content, we can pierce the dominant narrative that the animals being raised on farms are happy and that their deaths are quick and painless.

Arun is trying to figure out how to work with more communities of color and LGBTQ+ groups, finding areas where their work overlaps. What they have in common may be concerns about the climate impacts of animal agriculture or an opposition to the use of gas chambers at pig slaughterhouses. "It's not what you care about, it's what they care about; it's about finding common ground," says Arun, echoing Saul Alinsky—American community organizer, author, and mentor to Hillary Clinton—when he points to "the fundamental idea that one communicates with the experience of her audience." Alinsky further notes: "As an organizer, I start from where the world is as it is, not as I would like it to be. That we accept the world as it is does not in any sense weaken our desire to change it into what we believe it should be—it is necessary to begin where the world is if we are going to change it to what we think it should be. That means working in the system. . . . People don't like to step abruptly out of the security of familiar experience; they need a bridge to cross from their own experience to a new way."[22]

Redefining the Wins

To build a mass movement, we must have an approach that is welcoming and inclusive, but that's not enough to inspire and retain activists. Arun articulates a third requirement: "We've got to be successful. You have to

22 Saul Alinsky, *Reveille for Radicals*. Chicago: University of Chicago Press, 1946, pp. xviii, xix, and xxi.

have a purpose and measurable success. We have nice social activities with activists. We have a community. That community must see signs of progress and hope." That's what keeps Arun and Silvia going week after week, spending much of their own time and energy to stand in the cold at early-morning vigils. "You do that if you feel you can make a difference. We don't lose heart because we can see progress."

For our movement, one of the most important ways to "win" is to increase the number of people bearing witness. Bearing witness is life-changing and transformative, but it is incredibly difficult to do and most people, including vegans, don't choose to do it. Alex Lockwood, a media studies instructor and author of a book about how bearing witness changed him, who worked closely with Manchester Pig Save, says: "Our culture and society don't prepare us well for hard, emotional practices. That's an element. We are struggling against strong and sustained scripts of what we can do." The question is: Are there creative ways of exposing people to this form of activism without seeming to force them to bear witness?

One innovative way to secure some small wins and get new groups of people to bear witness was through the Manchester Vegan Run. In 2016, Alex, a runner, originally proposed the run as a way to bring the constituency of vegan runners to a vigil by getting them to run around the perimeter of Tulip slaughterhouse and "accidentally bear witness." "By letting them do something they were comfortable with doing—running—we get them to do something they are uncomfortable with—witnessing animals at the slaughterhouse." Also part of the event were talks and free food, including Gregg's sausage rolls. The tradition continues. The 2022 run organized by Arun and Silvia attracted 150 runners and raised £2,000 for local farmed-animal sanctuaries. "We have to empower others," says Arun, "otherwise rates of change slow if you do everything yourself."

Jane Velez-Mitchell says no one would want to bear witness unless there's a benefit, saying: "I just took a leap of faith. There is a benefit, even if it makes me suffer. I'll be honest, I hid behind the camera a lot because it was so horrible. When you're a journalist, recording automatically makes you more detached. When I looked at the animals and looked in their eyes, it was devastating. I am ashamed of being part of the human species. The monumental scale of animals being killed is almost too much to bear!" So, what is the benefit? "When you bear witness, there is power in that. Pigs appreciate the water, gentle touch, and love—as they go into the most horrific experience. But when you bear witness as a group, it's exponentially beneficial. It says to the pig, 'You count, you weren't forgotten, you have value—you count in the universe.'"

An extra benefit is that the vigils reach the wider public, outside the vegan bubble, thanks to footage shared on social media and in mass media. Footage of frightened and sad pigs frothing at the mouth is worth a thousand words. "Once you see it, you can't unsee it," says Jane. Amanda Holly, vigil organizer in Brisbane, Australia, draws attention to how vigil images have transformed the public's perception of chickens. "People have an image in their head that chickens are free-range. Then they see the wretched animals on the trucks, with burns from ammonia and their legs splayed. It drives home that everything is not what the industry says it is."

Another "win" is having celebrities attend vigils. Prior to the pandemic and when the transport trucks used to stop in front of the gates of Farmer John pig slaughterhouse, LA Animal Save vigils were the largest worldwide, drawing 150 people each week. The pig vigils regularly attracted A-list celebrities, including Joaquin Phoenix, Rooney Mara, Moby, Kat Von D, Mena Suvari, Craig Robinson, Damien Mander, Toby Morse, and Nimai Delgado. Organizer Amy Jean Davis says: "I don't think anything super sparked it, but I know that with Shaun [Monson, director of the documentary *Earthlings*] and Joaquin coming, it drew some people in. But honestly, I think it's mostly the ability to give water that gets people there!" Jane Velez-Mitchell's team was there to livestream when Joaquin Phoenix arrived at the pig vigil, still wearing his tuxedo after he had won the Screen Actors Guild's award for best actor in 2020. This was covered by *People*, *ET Canada*, and other mainstream magazines.

Another way of attracting more people to vigils is to hold special, all-day vigils, which Toronto Pig Save has done many times, usually in the warmer months when the attendance would be much higher. These vigils would combine vegan outreach in front of the slaughterhouse or at a nearby intersection, vegan food giveaways, special guest appearances, workshops, music performances, chalktivism, and artwork displays. Our largest all-day vigil was at Fearmans pig slaughterhouse on September 24, 2015, when we invited special guests Bite Size Vegan and Amy Jean Davis. (Amy would set up LA Animal Save upon her return home.) Four hundred people came throughout the day, with more than half being new attendees who had never borne witness before. Emily of Bite Size Vegan made a profound video entitled "Open Your Eyes" from that day. The powerful effect of so many people bearing witness brought the manager of Maple Leaf Poultry—one of the slaughterhouses in Toronto—to tears. He surrendered a chicken from the kill floor whom activists named Mercy and who was then able to live out the rest of her life at a local sanctuary.

In Australia, legendary organizer Patty Mark and her team at Animal Liberation Victoria organized a five-day vigil and fast in 2017. Manchester Pig Save followed suit with a five-day vigil in front of Tulip pig slaughterhouse. In June 2018, Dusseldorf Animal Save held a five-day vigil, called Animal Justice Camp, in front of Tönnies slaughterhouse, Europe's biggest pig slaughterhouse, in Rheda-Wiedenbrück in North Rhine-Westphalia, Germany, which kills 25,000 pigs per day and 20.6 million per year. The Barcelona Save chapter added a five-day fasting component to its special vigils in 2018 and 2019.

5

The Killing of Regan Russell

THERE WAS A LONG HISTORY of safety issues that activists ran into with Sofina Foods' Fearmans pig slaughterhouse in Burlington, Ontario. We met with the police on two occasions to attempt to negotiate safety agreements. The meetings went nowhere. So we drafted a petition directed towards the CEO and owner of Sofina Foods himself, Michael Latifi. The change.org petition, "Michael Latifi: Stop Endangering Activists! Negotiate a Safety Agreement at Fearmans Now!" garnered over ten thousand signatures. We wrote to Mr. Latifi, and the only reply we received was from his lawyers, warning us not to write to him again. In 2019, during an animal rights conference in Toronto, we held a demonstration in front of Latifi's home in the Bridal Path area of Toronto. We staged a "die-in" on his front lawn and posted placards with photos of individual pigs who had been killed at Fearmans.

It is very difficult to bear witness and connect with the animals unless the trucks stop for at least a few minutes. At the cow and chicken slaughterhouses in Toronto, activists initially had to stop the trucks by standing in front of the entrance. Early on, Toronto Chicken Save negotiated with the plant manager at Maple Leaf Poultry so that at the weekly vigils, each transport truck would stop for five minutes. The agreement holds to this day without incident. Similar arrangements with the cow slaughterhouse have been made on and off. The security guard at Ryding-Regency cow slaughterhouse kept the gates shut to the arriving transport trucks for a set time for activists to safely bear witness.

Other chapters around the world have also achieved safety agreements. For example, LA Animal Save worked with the Vernon Police Department to stop each truck for a couple of minutes at Farmer John pig slaughterhouse. Lead organizer Amy Jean Davis told me: "We are fortunate to work with the amazing Vernon Police Department. . . . The way it works is we have people who will stand in front of the gate on the sidewalk when a truck is approaching. This is what stops the trucks." LA Animal Save held the largest vigils in the world before COVID-19 struck, often with 150 people attending. At the vigils, Amy reads out safety guidelines to maintain order: "Always approach the trucks slowly, calmly, and with a gentle voice. Water is the priority; photos and videos are secondary. Work with a buddy if you can, one giving water and one documenting."

Friday, June 19, 2020

On Friday, June 19, 2020, time stopped. Regan Russell, a pioneer in animal rights activism, was violently struck and killed by a transport truck filled with pigs in front of Fearmans slaughterhouse. She was at a Toronto Pig Save vigil with six other activists, giving water to pigs on one of the hottest days of the year. Regan regularly attended pig vigils and on this particular day, she was also opposing Ontario's "ag-gag" Bill 156, which had passed the day before, on June 18. The bill criminalizes whistleblowers, journalists, and activists who play the vital role of exposing the horrific treatment of farmed animals. It specifically targets groups, such as Animal Save Movement, that peacefully bear witness outside of farmed-animal slaughterhouses, provide water to dehydrated animals, and document their suffering.

Regan was run over while crossing on a green traffic light—she had right of way—to join her fellow activists at the end of the vigil. Her death was an enormous shock and a huge loss to not only the Canadian but the global animal rights community.

In the aftermath of Regan's killing, there was a collective coming together. People felt the same emotion, no matter where in the world they were. At that moment, nothing else mattered. Animal Save Movement immediately organized global events for Regan; we were determined to not let Regan die in vain. People all over the world grieved and held memorial vigils, marches, and demonstrations in Regan's honor.

Regan was deeply disturbed by ag-gag Bill 156. Her last post on Facebook, published the night before she was killed, decried it as a dangerous piece of legislation. Regan was determined to stop this bill. Before her death, she had messaged Lori Croonen, a local vigil coordinator, to arrange a meeting with Lori and our team to strategize on how to defeat it. She had stated that she and her ninety-year-old father Bill were willing to go to jail to challenge it.

Tragically, the new ag-gag law increased the risk of dangerous driving by slaughterhouse truckers, since it offers farmers protection against civil liability for anyone hurt while contravening the act. Our fear that the provision would enable and embolden transport truck operators to drive aggressively, endangering the lives of peaceful and law-abiding citizens, was realized just one day after it was passed. Mark Powell, Regan's partner, who brings her own fluorescent, bright, clear signs with big writing to the courthouse, says: "This is a fight for the sanctity of our rights as free citizens to demonstrate for what we believe in. We seek to repeal Bill 156 in its entirety, and let it be known we believe Regan's blood is on the hands of those who supported this vile legislation."

Animal Justice, a group of animal rights lawyers in Canada, is challenging the constitutionality of Bill 156. Regan was struck while holding a water sprayer, which she had used earlier that morning to give water to thirsty pigs. It was her final act of kindness. She had been standing up for the right to bear witness. She had been exemplary in her love-based activism. That morning, just minutes before she was killed, another truck driver, also from Brussels truck company, had come to a complete stop at the slaughterhouse gates to let activists safely bear witness to the pigs. Russell had gone up to the passenger side, looked the driver in the eye, and said, "Thank you for stopping."

Over four decades, Regan was arrested eleven times for various acts of nonviolent civil disobedience on behalf of oppressed humans and nonhuman animals. Regan famously said: "People say we're breaking the law. . . . How do you think women got the right to vote? How do you think slavery was abolished? People stood up and broke the laws—because they were stupid laws." Regan had marched with Black Lives Matter in Hamilton, her hometown, the weekend prior. She fought for Indigenous rights, for LGBTQ+ equality, and of course, for animal liberation.

Her untimely death sent shockwaves throughout the global animal rights community. Activists everywhere tried to turn their sorrow into action. The Toronto, Hamilton, and Burlington communities launched a Justice For Regan pressure campaign, with marches in these cities and a demonstration at the Halton police station. We called for a transparent criminal investigation and the laying of appropriate charges regarding Regan's killing. Regan's family joined in, demanding that ag-gag laws be repealed and the right to bear witness be guaranteed. Mayor Marianne Meed Ward dedicated November 1st, or World Vegan Day, in 2020 in the City of Burlington to Regan, to celebrate those who raise awareness about the treatment of nonhuman animals and to recognize the benefits of plant-based diets.

Around the globe, activists held memorial vigils and organized direct actions and banner drops in 120 locations. In Berlin, activists occupied a chicken slaughterhouse to demand a transition to a healthy, sustainable food system. A Regan Russell banner was hung from a rooftop as part of the action, which saw peaceful activists carried away by police. In India, Animal Save Movement teamed up with Million Dollar Vegan India to distribute delicious vegan meals to 4,000 people, including migrant workers, in the Dharavi Mumbai India slums as part of the Food Justice For Regan initiative. In Los Angeles, actor Joaquin Phoenix joined a pig vigil in honor of Regan and held a sign that read #SavePigs4Regan. He later attended a pig vigil at Fearmans slaughterhouse in Burlington and stood at the spot where Regan was killed to honor her and to use his influence to speak up for all the animals slaughtered: "Regan Russell spent the final moments of her life providing comfort to pigs who had never experienced the touch of a kind hand." In the summer of 2020, the Portuguese Parliament unanimously approved a vote of sorrow in honor of Regan. Across Europe, the Grand Prix was targeted due to Sofina owner Michael Latifi's investment in the McLaren and Williams racing teams. In the UK, activists glued themselves to the streets outside the Canadian embassy in London, blocking the road to call for the repeal of Bill 156 and for justice for Regan.

The twenty-eight-year-old who drove the Brussels Transport Ltd. truck that killed Regan has been charged with merely careless driving causing death—a provincial Highway Traffics Act infraction, not a Criminal Code offense. This feeble charge does not amount to justice for Regan. In a statement, her stepson Josh Powell said, "The lax charge in response to her death has only made the pain deeper."

Regan Lives on and on and on

In tribute to Regan, activists and farmed-animal sanctuaries rescued dozens of chickens, pigs, lambs, and goats. Many animals continue to be rescued in her name. Days after Regan's death, PETA rescued two six-week-old piglets from a farm in Iowa, naming one Regan and the other Russell. These survivors of the animal agriculture industry are now living at Arthur's Acres Animal Sanctuary in New York, where they will be free to frolic, wallow, forage, and enjoy the companionship of other pigs. Little Bear Sanctuary in Florida rescued Regan the pig, along with her companion Rosie. Black Sheep Sanctuary in Cancun, Mexico, and Ledwich Animal Farm in Manitoba, Canada, both rescued a pig. In France, DxE Paris rescued a goat and her two babies and Paris Animal Save rescued a chicken. In the UK, fifty three-week-old turkey poults being reared for Christmas were spirited away to freedom in Regan's honor, to be given sanctuary and a life free from violence and oppression.

Lori Croonen, who had been there at Regan's last vigil, wanted to do something to help farmed-animal sanctuaries in Regan's name. We set up a planning committee. We launched an Indiegogo crowdfunder with perks like T-shirts, hoodies, and animal rights books for the first anniversary of Regan's passing and raised over US$10,000, which were distributed to twenty-four sanctuaries.

On the one-year anniversary of Regan's death, the Ollerton 11 hit UK headlines after a pregnant pig named Matilda escaped from a factory farm in Nottinghamshire. Matilda's motherly instincts kicked in and she gave birth to ten piglets in a nearby woodland. More than 6,000 people signed a petition demanding that Brinsley Animal Rescue be allowed to save the pigs. The farm reluctantly agreed to hand over the family and spare them the horrors of slaughter, on the condition that a planned protest was called off. In a Facebook post, Brinsley Animal Rescue said: "The first piglet to be named is now Regan, in memory of the brave Regan Russell, who tragically died a year ago today whilst protesting at a pig slaughterhouse."

Regan once described herself as a mere pebble on the road to freedom, but she left behind a legacy of love and compassion and continues to inspire activism worldwide. Many new activists arose as a result of Regan's tragic killing to continue her mission to protect all of humankind's animal victims on this Earth. Her death increased the resolve of many activists to show up on a regular basis and be there for the animals at slaughterhouses. Many of these people had never attended an animal vigil before and only became involved specifically because of Regan, especially in the first year following her death. There was an incredible sense of collectiveness.

Regan once said: "I don't know if [activism] does any good. But I know doing nothing does no good." She has inspired many activists around the world to do a whole lot of good for the animals and the planet. Regan ignited a fire in a number of incredible activists who became core Save organizers in her name, determined to propel the movement for animal liberation forward. Nicola Harris in the UK, Adrienne George in

Welland, Ontario, and Varun Virlan and Peter McQueen in Toronto all became much more involved because of her.

Nicola Harris is someone I admire greatly. I don't have words to describe her compassion or dedication, or the exceptional skills she brings to helping animals and developing creative and impactful campaigns. The first Stop Huntington Animal Cruelty campaign (SHAC UK)—in which Nicole was involved—was so successful that the authorities set up a corrupt operation to shut it down by unjustly imprisoning Nicola and the other young animal activists for almost a year.

I first met Nicola during the extremely traumatic time immediately following Regan's death. Nicola had read on Facebook that someone had been killed at a slaughterhouse vigil. She immediately thought of Jill Phipps, a British animal rights activist run over by a slaughter truck at Coventry airport on February 1, 1995, while protesting the live export of baby calves, not to mention the several hunt saboteurs who had also been run over. "If you're in the UK, it really affects you," she said. She was in tears and promptly did an online search, wanting more information. "Once I got Regan Russell's name, it made it more personal. I wanted to know more about her and who she was. I went to her Facebook page. I saw she was this really beautiful soul. I was so moved that she had stood up for something and lost her life for it. I was so upset for her family, her friends, and her community. I was thinking that in the UK, when protesters are killed in protests, there's never any charges, never any justice."

For example, no charges were brought in the case of two hunt saboteurs run over by hunting vehicles, despite there being witnesses. Mike Hill, the first hunt saboteur to be killed, died in 1991. He was eighteen years old and animal rights activists believed he was deliberately run over. Nicola says that police are often part of the hunting community and of that social class. And while the police say they didn't see any crime, statements from the animal rights community are seen as not objective. Tom Worby was only fifteen years old and at his first hunt sabotage when he was killed in 1993. Nicola says: "Not that charges bring justice. But it's an extra injustice when someone stands up for something and the authorities don't take it seriously."

"It really lit something inside of me. This woman had lost her life to create this vegan world and save animals. I'm so lucky to be alive. I have the duty to carry on her work because she can't do it now. We all have to do that now because she's part of our community. At that point, I had been struggling with reintegrating into the animal rights movement. After I got out of prison, it was really difficult to trust people. I was doing little bits for animal rights. I was struggling. . . . This helped me overcome it. Take the risk and just go for it, really."

Before Regan's death, Nicola had attended a couple of vigils with Dorset Animal Save, the Save group closest to where she lived. It was her first time attending an animal rights event since she had been incarcerated. "I found it quite difficult when people were trying to make friends. I didn't know if they were police. It was causing paranoia. . . . I met these beautiful cows. They were high up in the trucks. I could see their little noses." She was also undertaking investigations at chicken farms and was filming in pig farms for Viva!, a prominent animal rights group. "I've seen all the different animals. People say billions of land animals are killed each year, but when you meet them one-on-one, you just feel this duty to fight for them."

Nicola joined Animal Save Movement by chance. She had been speaking with Gerrah Selby, who had also been part of SHAC years earlier and now worked as a graphic designer for us. She suggested we meet. Nicola had been the communications lead for SHAC when she was arrested a decade ago. We immediately recruited Nicola to join our communications team, which was focusing on garnering public and media attention to get justice for Regan and carry on her mission.

Adrienne George didn't know what a vigil was before Regan's death. She was too nervous to join protests, especially if she didn't know anybody or where they were held. After going vegan, she became active with the Animal Alliance of Canada, doing local activism and meeting with Members of Parliament to advocate an end to cosmetic testing on animals.

She heard of the protests at Fearmans slaughterhouse, but the barriers to participating were too large for her: "In my mind, a slaughterhouse is out in the middle of nowhere. It's the fear of the unknown—of not knowing anyone. I didn't know it was next to a Tim Hortons coffee shop."

When Adrienne heard about a memorial for Regan, she knew she had to attend. "I had a day off, so I could go. If Regan died for her beliefs, I can get off the couch and go to the frontlines." As she parked at the GO (commuter train) station next to the slaughterhouse on a Sunday, she was taken aback by the mundaneness of the scene: the Tim Hortons across the street, a gas station, and "people driving by like it [was] everyday business." It wasn't the middle of nowhere after all. At the memorial, Adrienne heard about the pig vigils on Wednesdays. "Regan's death introduced me to who I am now. I went to my first vigil and haven't stopped coming. She opened the doorway." At vigils, Adrienne talks to and makes eye contact with the pigs. She sometimes goes up to the chain-link fence in front of Fearmans and tries to imagine what is happening to the pigs behind those corrugated, gray walls. "Yet, people are driving by with their Timmies coffee and donuts, not knowing 10,000 pigs are being killed while they have a ham sandwich in their hands that they'll forget about in ten minutes. That didn't depress and immobilize me; it lit a fire instead."

Adrienne says she found "her people"—activists with the same values and beliefs—that week at the memorial and pig vigil. "I was really drawn to Save and the idea of returning negativity with love and respect. I wish that I had someone tell me a lot earlier what I know now and that it didn't take Regan's death to get me out of my comfort zone." Adrienne had twenty years of experience in marketing, having worked for large companies, such as Tim Hortons, and companies in the music industry. After joining Toronto Pig Save, Adrienne immediately took on marketing initiatives, creating, for example, a Calendar for Regan, with photos highlighting Regan's activism. Miraculously, she got Canada Post to issue a postage stamp with the words "Go Vegan for Regan." She knew Canada Post had a service that offered customized government-issued postage stamps but wasn't sure they'd accept that caption. As it turned out, she just had to provide authorization from Regan's family to use Regan's image. Adrienne also organized a six-hour vigil (which was supposed to be a twelve-hour vigil but was trimmed down because of COVID-19) and raised hundreds of dollars to donate to Animal Justice. At our Toronto march for Regan, she was our media point person at Queen's Park, the Ontario Legislative Building, greeting the media and running the table with outreach material.

Several documentaries were made for Regan. Michele Labrosse interviewed Regan in depth in front of Fearmans in the winter a few months before she was killed. When Michelle learned of Regan's death, she immediately went to the slaughterhouse for the vigil. Upon returning home, she worked on the video and released it on the same day on her YouTube channel. She wanted the world to know who Regan was. The video helped to make known Regan's long history of activism and to rally people to seek justice and continue Regan's legacy. A few days after Regan's passing, Varun Virlan also made a truly moving and poignant documentary about her, using Michelle's interview along with other activists' footage. He called it "Regan Russell: A Short Film." Varun said: "We felt it was important for people around the world to know the story of Regan Russell. She was a friend, fellow activist, and inspiration to many. Since her death, we made a commitment that she would not die in vain."

Many documentaries on Regan have drawn on videos of her taken by Agnes Cseke—from vigils, protests, and the Pig Trial. Agnes is a self-described "obsessive documentarian" who records the animals and all the activists at vigils. She says: "I didn't know Regan. We said 'good morning' a few times to each other but I didn't have the chance to really have a relationship with her. Mostly, I was shy to video-record people that I didn't know yet, and I felt that I was sort of stealing when focusing on her, but I did it anyway with discretion. I think that I have in my videos every activist that was present at the vigils that I went to. It just makes sense to me to concentrate on the full event. I consider it history in the making."

Camille Savage, organizer of Las Vegas Animal Save, always has signs in honor of Regan at the chapter's monthly chicken vigils. For the second anniversary of Regan's death, the group planted a memorial tree for Regan at All Friends Animal Sanctuary. The tree and mulch were donated by Nevada Plants, a group that focuses on bringing plants into urban areas. *There Was a Killing*, a documentary by *Earthlings* director Shawn Monson, was screened at the memorial. The documentary tells the powerful story of Regan and her worldwide impact, featuring firsthand accounts and in-depth analyses from attorneys Robert Monson (Shaun's dad) and Lisa Bloom. Shaun says, "In exploring the story, including a new animal agriculture-sponsored law that appeared to be a license to kill, one can't help but wonder if there was a cover-up and corruption involved." Regan Russell's martyrdom was also covered in a documentary for *Now This*, which has garnered over six million views on Instagram and Facebook.

British singer-songwriter Barbara Helen wrote a memorable song for Regan called "One Voice for Regan," for which Shaun Monson created a poignant music video featuring archive footage of Regan Russell at vigils and protests. Lyrics include: "Where there was darkness, you were the light, shining so strong and shining so bright. We will keep fighting and your memory will go on and on."

Regan's Principles

Following Regan's death, Peter McQueen became lead safety marshal, along with Adrienne, at Toronto Pig Save vigils. Peter learned of Regan's death from Lori's message in a Toronto Pig Save Messenger group chat: "Regan was killed. She was run over and cut in half at a vigil at Fearmans." It was shocking. Peter had met Regan a few times at various activist events.

Peter says he organizes photos on his phone by people's names, but in Regan's case, he could filter her photos thanks to her fluorescent signs, like her bright orange sign that read We Are Their Voice at a Mother's Day march from downtown Burlington to Fearmans on May 19, 2019. Regan always had good signs. She wrote memorable sayings like All Animals

Need Protection under the Law and If You Were in this Truck, We'd Be Here for You Too.

Following Regan's death, we put more stringent safety rules in place for vigil participants. We created official marshal positions and insisted that everyone wear safety vests and follow a host of rules. To stop trucks in a much safer way, we only used caution tape held by two activists across the entrance of the facility. Peter says: "It felt good being part of a team dedicated to helping keep people safe. A number of trucks did stop and allow us to bear witness this way. By doing it this way, we are being not only safer but also less confrontational. Part of what Adrienne and I do is setting a tone and setting expectations for people. It's important to set the ground rules and make clear the philosophy of love-based organizing. I find, by and large, that people respect the approach."

Adrienne adds: "Peter and I take the role very seriously. We've developed good relationships with the police—they arrive, they say, 'Just be safe.' Our main goal is to keep activists safe and to have an experience that keeps bringing activists back. Peter tends to stay on the sidewalk. I tend to go on the traffic median, where the bearing witness occurs. Peter will greet new activists, give them a rundown, and hand out safety vests. I tell them what to expect when the truck comes. . . . I make sure they know what to expect and make them feel welcome by introducing them to the other activists. I can't even tell you how amazing it is to be with other activists, even when you've just met them. I am grateful to be with them . . . like I said, we're a family."

We also developed what we called the Regan Principles, which we read before our Toronto Pig Save vigils:

Respect everyone, including the animals, activists, the public, and employees of the industry.
Every interaction with the public needs to be positive. They just haven't come around yet. You are an ambassador for a new, just, and nonviolent vegan world.

Give everyone the opportunity to hear, and learn, about what we are doing.
Always be disciplined and take care of each other.
Now that Regan is gone, may her teachings be your guide. Take people under your wings and bring them to bear witness.

It has been moving to learn of how Bill, Regan's elderly father and a former school principal, had joined her in animal rights protests since 1987 and became vegan with Regan as his "best teacher." Regan's parents join the pig vigils in her memory and find great solace in the outpouring of support from activists around the world continuing the fight against all oppression that was central to Regan's life.

Following Regan's death, the conservative Ford government of Ontario began passing regulations to implement the ag-gag bill. Previously, the police had not been enforcing the legislation. The first regulation would prohibit people from blocking a truck or, as the government called it, "hindering a load." Peter asks rhetorically: "A load of what? Are they pigs or pig iron?" The measure had a major impact on our ability to bear witness at vigils. The next regulation to be introduced would prohibit people from interacting with the pigs, including giving them water, or "interfering with the load." "Our giving water to pigs on a hot day gave the pigs some respite and I'm sure they felt the love. For a lot of people, it's frustrating not to legally be able to give the water. Maybe that's why some people stopped attending the vigils, feeling they weren't being much help."

We had to adapt and change tactics. Peter says he didn't find it effective to be at the entrance of Fearmans slaughterhouse. "So, we went back to the narrow traffic median." There, the trucks sometimes stop at the red light before turning towards the slaughterhouse. This enabled activists to bear witness, often for just fleeting moments, and to be able to take pictures. They could also do outreach at the busy intersection, holding signs and interacting with drivers. Peter says interactions at the meridian were better because at the previous location, the trucks would often purposely block traffic and it upset a lot of drivers. He shares that now, drivers offer "more positive comments, a thumbs-up, a peace sign, or a friendly honk." "It happens at every vigil. Even if we get negative comments, it gives us a chance to interact. It's the people yelling 'bacon' that it's hard to interact with. I give them a peace sign." That's what Regan would have done. She would have flashed her fluorescent sign, gazed at the yelling person with her warm, sunshine smile, and said: "We are here for the animals. If you were in that truck, we'd be here for you too. Join us!"

6

How to Be a Good Organizer and Sustainable Activist

"Do not believe in words, yours or others'; believe in the deeds," said Tolstoy.[23] He believed that to act morally is not just to think good thoughts but to take action. To grow and flourish, a grassroots movement requires a plentiful supply of good organizers and new recruits. Consider that the civil rights movement in the US had hundreds of professional organizers, tens of thousands of organizers, and hundreds of thousands of activists working to achieve freedom and equality.[24] Cesar Chavez, founder of United Farm Workers, said, as cited on the organization's website: "A good organizer has to work hard and long. There are no shortcuts. You just keep talking to people, working with them, sharing, exchanging and they come along."

Self-Purification

For Mahatma Gandhi, the starting point for activism was self-purification, involving prayer, fasting, self-discipline, and a redefining of one's roles. Like Leo Tolstoy, he believed that only an individual whose soul is pure

23 Tolstoy, *A Calendar of Wisdom* [December 13], p. 360.

24 Eric Mann, *Playbook for Progressives: 16 Qualities of the Successful Organizer*. Boston: Beacon Press, 2011.

may begin to consider others in their surroundings. "To see the universal and all-pervading Spirit of Truth face to face, one must be able to love the meanest of creation as oneself. . . . Identification with everything that lives is impossible without self-purification. . . . God can never be realized by one who is not pure at heart. Self-purification therefore must mean purification in all walks of life. . . . Purification of oneself necessarily leads to the purification of one's surroundings."[25] Gandhi's autobiography documents his starkly honest and profoundly self-critical "experiments with truth": "I cannot claim success for any experiment. . . . My object is only to show that he who would go in for novel experiments must begin with himself. That leads to a quicker discovery of truth, and God always protects the honest experiment."[26]

Gandhi proposed that the individual who is willing to admit their own imperfections and fallibility in their journey is more likely to find truth and to have the capacity to improve the social conditions in which they live. In his personal life, Gandhi experimented with the legal profession, journalism, diet, exercise, simple living, religion, child rearing, and abstinence. For example, he redefined the task of a lawyer. As a lawyer himself, he realized his loyalty did not rest first and foremost with his client but rather with his "devotion to truth." In a case in which Gandhi discovered that his witness had been dishonest, he realized he had a false case and told the magistrate to reject the case. Gandhi affirmed his belief that "it was not impossible to practice law without compromising the truth."[27]

As a life-long journalist, Gandhi also advanced a definition of social movement journalism, declaring that "the sole aim of journalism should be service" to the community.[28] He saw journalism as playing an essential part of social movements, providing not only accuracy in news coverage but also remedies from a social movement viewpoint to end

25 Mahatma Gandhi, *An Autobiography: The Story of My Experiments with Truth*. Ahmedabad: Navajivan, 1940, p. 504.

26 Ibid., p. 307.

27 Ibid., p. 366.

28 See S. N. Bhattacharyya, *Mahatma Gandhi: The Journalist*, London: Asia Publishing House, 1965.

discrimination, achieve political emancipation (Gandhi's newspapers asked readers to "literally take their weekly lessons in nonviolence"), elevate mass consciousness (his newspapers also included book reviews from a service perspective and biographies of great figures such as Tolstoy), and encourage self-critical thinking.

It's important for activists and organizers to have a mindset that prioritizes self-reflection and self-improvement. They must also realize that community organizing doesn't necessarily happen automatically and easily: they are in it for the long haul. Organizing often involves difficulty and drudgery. Alinsky says in *Reveille for Radicals*: "It can be done only by setting ourselves to the dirty, monotonous, heart-breaking job of building 'Peoples' Organizations.' It can be done only by possessing the infinite patience and faith to hang on as parts of the organization disintegrate; to rebuild, add on and continue to build."[29]

In addition, there needs to be constant training and development of leadership skills, including knowledge and communication skills. "Organization doesn't come out of an immaculate conception. It takes a highly trained, politically sophisticated, creative organizer to do the job."[30] Alinsky believed so strongly in the importance of having trained organizers that he set up a training institute in Chicago in 1968. In the book *This Is an Uprising*, mass training is identified as critical for grassroots organizing because it enables movements to absorb momentum and grow exponentially. It involves training the trainers. Initiation of mass training for new members of any given movement is essential in establishing its grand strategy, theory of change, campaigns, and tactics, and in developing the DNA of the organization. That is what is required if the movement is going to win. Our slaughterhouse vigils, made possibly by love–based community organizing, serve as a template to help people plan replicable actions within the DNA characterized by nonviolence, nonjudgement, and the principle of returning love for hate.

---

29 Alinsky, *Reveille for Radicals*, p. 219.

30 Marion K. Sanders and Saul Alinsky, *The Professional Radical: Conversations with Saul Alinsky* (Evanston and London: Harper and Row: 1970), p. 68.

Love-Based Activism

One reason people keep coming back to our animal vigils is because they are love-based, meaning that anger and confrontation are strongly discouraged. Instead, organizers and all attendees are urged to remain calm and exude compassion and kindness, to be grateful to everyone who attends and helps out, and to introduce people to each other and thank them for joining. I've learned how important it is to constantly work on the love-based philosophy of community organizing. The challenges of trying to save animals and the planet are so immense and urgent, it is easy for us to fall into the trap of feeling discouraged and angry—at not only the system of animal agriculture but also the individuals participating in it.

I find it helpful to reread, once in a while, works by Leo Tolstoy, Mahatma Gandhi, Martin Luther King, Lois Gibbs, Cesar Chavez, and other practitioners of love-based organizing and communication. Tolstoy liked to quote the biblical passage, "I came not to judge, but to save." He advocated returning love for hate, kindly pointing out to the oppressor what is wrong, and organizing the public through "noncooperation with evil" campaigns based on methods such as boycott and direct action, including nonviolent civil disobedience.[31]

Historically, community organizers have recognized the importance of activists having love and respect for people. Alinsky says: "The Radical does what he does because of his love for his fellow people."[32] This is challenging for many animal rights activists who identify more with oppressed members of other animal species, but it is an essential characteristic to possess if we hope to persuade others to change their diets and join the animal

---

31 These themes are present throughout Tolstoy's nonfiction works, including *A Confession*, translated by Aylmer Maude (Mineola, New York: Dover Publications, 2005, first published in Russia in 1882); *On Life and Essays on Religion*, translated by Aylmer Maude (London: Oxford University Press, first published in 1887); his essays between 1894 and 1909 in *My Religion: What I Believe*, translated by Huntington Smith (Guildford: White Crow Books, 2010, originally published in 1884); *What to do?*, translated by Isabel F. Hapgood (Aegypan Press, Thomas Y. Crowell edition, 1887); *The Gospel in Brief*, translated by Isabel Hapgood (Mineola, New York: Dover Publications, 2008); and *The Kingdom of God Is within You: Christianity Not as a Mystic Religion but as a New Theory of Life*, translated by Constance Garnett (New York: 1894).

32 Saul Alinsky, *Reveille for Radicals*. Chicago: University of Chicago Press, 1946, p. 113.

rights movement. Saul Alinsky writes: "An organizer has to have respect for people and their program. . . . An organizer who has this superior attitude cannot, in spite of all his cleverness, all his protestations of belief in the equality of all people, including himself, conceal his true attitude. It repeatedly comes out in a gesture, an expression, or the inflection of his voice. . . . An organizer who really likes people will instinctively respect them."[33] One way we can nurture this attitude is to work on active listening and approaching others from a place of love. People are much more responsive if we engage them in conversation with genuine warmth and a desire to understand them. In trying to recruit individuals and groups to a movement, we might heed what Tim Sampson, a community organizer, says, "the flowers of organizational relationships grow from personal interest, kindness, and cultivation."[34]

Gratitude for any help we are able to offer to the suffering animals and any help we receive from other activists in the movement is a step on the spiritually enlightened path and a key to movement building. Tolstoy, in his book *A Calendar of Wisdom*, quotes John Ruskin: "When you do a good deed, be grateful that you have had the chance to do it."[35] Successful community organizing requires strong and wise people to use their gifts to support and help each other to reach their full potential as activists and organizers.

Good leaders inspire others to lead, who, in turn, pass it forward and nurture others to spread the movement. The best leader can be seen as one who develops leadership in others. Many great social movement leaders, such as Gandhi and Cesar Chavez, devoted their whole lives to using bottom-up, grassroots mobilization to encourage and inspire leadership in as many people as possible. The aim of grassroots movements is to develop a culture of organizing in which everyone is an organizer. Animal Save Movement promotes the creation of a new norm that recognizes the duty we all have to bear witness. We all also have a duty to seek

33 *Ibid.*, pp. 123–24.

34 Cited in Gibbs, p. 198.

35 Tolstoy, *A Calendar of Wisdom* [March 25], p. 154.

out slaughterhouses in our community, hold regular vigils and vegan outreach events, and build community.

A good organizer gets to know lots of people face to face. As Saul Alinsky writes: "You obviously cannot get everyone in the community to know everyone else on a personal, human basis, but you can get the hundreds of little local leaders to know each other on a human basis. . . . There are the Little Joes that have some thirty or forty followers apiece. Their attitudes significantly shape and determine the attitudes of their followers. These Little Joes [and Little Josies!] are usually totally ignored in all programs superimposed by well-meaning outside agencies, whether they be in the field of recreation or adult education."[36]

Making Organizing Part of Your Life and Becoming a Full-Time Activist

In his first biography, Gandhi writes that real change, wherein justice prevails over injustice, would arrive not in some "dim and distant future" but "within a measurable time, the measure being the measure of the effort we put forth," for "the real success lies in the effort itself."[37] It is imperative that we encourage people to commit to greater levels of activism—to make activism a regular and important part of their daily lives. Community organizing helps people to see that they have the power to effect social change, especially when taking collective action in their community.

Gandhi's devotion to public work grew to occupy more and more of his time: "If I found myself entirely absorbed in the service of the community, the reason behind it was my desire for self-realization. I had made the religion of service my own, as I felt that God could be realized only through service."[38]

Lois Gibbs, an American community organizer and environmental justice advocate, started out as a working-class, politically inactive housewife in Niagara Falls. Her awakening as an environmental justice

36 *Ibid.*, p. 176–77.

37 James. D. Hunt, *Gandhi in London*, Revised edition. New Delhi: Promilla, 1993, p. 118.

38 Mahatma Gandhi, *An Autobiography: The Story of My Experiments with Truth.* Ahmedabad: Navajivan, 1940, p. 132.

advocate occurred in June 1978, when she discovered that her son's school was built on top of a toxic waste dump that could be the cause of his illness. Gibbs had no formal education beyond high school; when she began work as an activist, she was even surprised to hear herself quoted in the media and not sounding "as dumb as [she] thought [she] was."[39] And when she went organizing door-to-door, she was surprised that not a single one was slammed in her face. Gibbs began to realize her role as an organizer: "I'm not the kind of person that does things like this. . . . It was all a new experience. It was exciting but also frightening. I wasn't sure what to do next or how to do it."[40]

In 1980, the Love Canal Homeowners' Association won an evacuation from their neighborhoods after Gibbs and other community residents organized a series of protests and exerted pressure on New York Governor Hugh Carey while he was running for re-election. Love Canal residents followed Carey everywhere, carrying signs and distributing fact sheets and press statements.[41] In 1981, Gibbs helped form the national Citizen's Clearinghouse on Hazardous Wastes, which aims to prevent the pollution caused by dioxin from medical and municipal waste incinerators.

Lois Gibbs developed a "Stages of Change" chart that looks at how organizers emerge from former apathy. Community organizing often begins with a "rude awakening" as people educate themselves about injustices in their community. Then there are feelings of betrayal and anger at the lack of government and corporate action. Those feelings need to be channeled towards constructive actions. Many Save organizers have also come from a place of apathy yet have been motivated to become active as they felt a sense of duty to shine a light on the injustice of animal exploitation. It helps to look for lessons from other social movements and people who can serve as role models and mentors. Next comes the

39 Sharon M. Livesey, "Organizing and Leading the Grassroots: An Interview with Lois Gibbs," *Organization and Environment*, vol. 16, no. 4 (December 2003), p. 503.

40 *Ibid.*, p. 38.

41 Lois Gibbs and the Citizen's Clearinghouse on Hazardous Wastes, *Dying from Dioxin: A Citizen's Guide to Reclaiming Our Health and Rebuilding Democracy*. Boston, MA: South End Press, 1995, p. 191.

realization of the power of communities organizing together to achieve victories step by step. According to Gibbs, the stages end in a lifelong commitment to social change as one "[accepts] the centrality of activism in [one's life]."[42]

Doug McAdam, in his book *Freedom Summer*, follows people who participated in the Freedom Summer in Mississippi in 1964 as part of the civil rights movement's drive to register voters. He discovers that for many, these profound experiences developed into continuous social engagement. Activism becomes how we identify ourselves. This can be a point of no return. A lifelong commitment is often made when an instance of an intense form of activism acts as a catalyst for an individual's self-realization. Once they realize that activism is essential in life, they put their talents in the service of a cause, as did Gandhi, Chavez, King, Gibbs, and millions of ordinary people.

Being a Good-Enough Activist

The eternal question, "Am I doing enough?" or, "Am I a good-enough activist?" can even prevent people from becoming active. Paul Loeb, an author on citizen empowerment, talks about the trap of the "perfect standard"—waiting to achieve a certain social and economic status before becoming socially involved or endlessly collecting facts on an issue before standing up and speaking out.[43] The idea is that we don't have to be saints or extraordinary people to become activists. Loeb advocates being a "good-enough activist" and realizing that change is often "the product of deliberate, incremental action." The important thing is to just start. By taking steps, however large or small, people begin to feel increasingly connected to a cause and activist community. As they learn more about the issues, they also develop a heightened sense of injustice and further comprehend the imperative to take action.

---

42 Gibbs, p. 151.

43 Paul Rogat Loeb, *Soul of a Citizen: Living with Conviction in a Cynical Time*. New York: St. Martin's Griffin, 1999. Loeb brings activism down to the personal level, allowing individuals to see how they can turn idealism into action and become part of the process of creating history.

Every bit of activism helps. Loeb deconstructs the hero theory of history, fueled by misleading historical accounts that recall only a few leaders of social movements and merely "mention the conclusions." The actual mechanisms and process of social change brought about by grassroots social movements involve millions of ordinary people taking incremental steps. Heroines often stand on the shoulders of their predecessors. For example, Rosa Parks's decision not to give up her seat and go to the back of the bus had historical precedents—other women in the South who had previously refused to give up their seats, all of whose acts were the lead-up to a culminating event whose time had come. Hers was not a whimsical move but a deliberate act of resistance based on extensive experience, knowledge, and training on social change methods, and made possible by support from community organizations like Highlander Folk School.

Within the animal rights movement and particularly at animal vigils, we frequently ponder the appropriate response to the prevalence of speciesism and the dominance and normalcy of exploitation of and violence towards nonhuman animals. How can we possibly respond adequately to the incredible scale and degree of injustice and suffering inflicted upon these innocent animals, especially when so many parties are complicit in perpetuating this industry—animal farmers, transport drivers, and slaughterhouse owners, managers, and workers on the supply side (all heavily subsidized by the government); and seemingly disconnected grocery shops, advertisers, and consumers on the demand side?

Chris Foott of Manchester Pig Save has this to say: "I always feel what I do is not enough because we can't save everyone—we can't do everything! But I think approaching vigils with anger won't help anyone. The other option we have—promoting love and compassion—is still the best approach." When we compare the handful of animals saved at vigils to the tens of billions of land animals and trillions of marine animals killed each year, it's difficult to keep our end goal in sight. An overwhelming majority of humankind's animal victims will be brutally killed and it's not an option, at this point in history, to save them all.

When we do the math, it's easy to wonder, "What difference do animal vigils make?" For example, the largest pig slaughterhouse in the Netherlands, located in Boxtel, kills 22,000 pigs a day. Animal Save Netherlands coordinator Lea Goodett says: "A few years back, when the vigils started, they were the biggest thing activism-wise. I felt I wasn't a good-enough activist. We then worked to remove that mental barrier. We were making our activism bigger than we should." Lea worked to reframe the perception of activism. "It's not all sad—it's also empowering to see all the people here together. . . . [Our activism] reached the community. They see that it's not that hard to go to an animal vigil. They can handle it." Lea would reach out to vigil-goers twice after the fact, right after the vigil and then a week later, to see if they were alright. The vigils became popular amongst the public and started to include activists' family members as well as nonvegans.

Lea says that of the people who started the vigils several years earlier, only ten are still active. She thinks that their burnout was the result of focusing on doing as many actions as possible and not taking care of themselves. She made it her mission to change this. "We do a lot of different events that make us happy."

Lea helped make Save Squares the most popular form of vegan-outreach street activism in the Netherlands, following in the footsteps of groups like Earthlings Experience and Anonymous for the Voiceless. Save Squares are now being organized in a dozen cities in the Netherlands and many other places around the world. A Save Square is made up of two teams of activists: One team stands in a square formation and, with blindfolds on, holds up informative signs or monitors showing footage from inside animal farms and slaughterhouses. The blindfolds are used to help bystanders come closer, without fear of being watched, and also represent society's blindness towards the various forms of discrimination faced by nonhuman animals. A second team hands out small information cards to interested members of the public while engaging in conversation using the Socratic method, which allows people to form their own opinions and reach their own conclusions regarding issues of animal, climate, and social justice. Save Squares are held in downtown areas with high foot traffic. There are now more Save Squares than vigils in the Netherlands. Lea says: "Lots of activists start with this type of activism and then get acquainted with our vigils—the core of our organization." At first, Save Squares were perceived as not very social in nature. Activists were asked to not talk to each other and to remain silent. Lea changed that: "It is a social event. . . . Activists are asked to be mindful, to make sure they don't miss the people walking by, but at the same time, the message is, 'Please talk to each other.' We want the experience to be fun."

The group began to organize many social, artistic, and musical events; recently, they even bought and painted a food truck. "I think we need this," says Lea. "For many, vigils are a bridge too far to cross. Street outreach is easier. Still, a lot of people don't want to stand with blindfolds in Save Squares either." Her recruitment strategy focuses on making it as easy as possible for people to do activism so that "anyone can do something with Save." Lea explains: "We need to come up with forms of activism that are even easier to engage with, such as activism involving food or art, or the Plant Based Treaty." The Plant Based Treaty is a solution-focused climate campaign with the motto "Eat Plants; Plant Trees." Instead of "just seeing sadness and cruelty" at slaughterhouses, we would do well to focus on climate action, too.

Lea's model is different from what I and other Toronto organizers had preached in the early years, namely that vigils should be frequent, preferably weekly (as is the case in Toronto). Lea says we need to focus on being a "good-enough activist" and changing the cultural demands. "A few vigils a year can be enough for some places, such as Denmark or Germany, where the whole country will be there. In the Netherlands, a lot of chapters hold vigils twice a year. The important thing is to make it happen. This takes away the pressure. It can be exhausting for some communities to do a vigil every week or every month." Lea is right. It's good to provide a menu of options. It's possible to have weekly events in some cities, mainly larger ones, but in towns where the animal rights community is not as developed, it may be more beneficial to hold fewer vigils and activist events—and make those events count. Footage of the animals from vigils can still be shared throughout the year on social media. Lea says: "We came up with a system where we tried to get that into people's heads—that everything they do is enough. We talked about this a lot. We need to get away from the idea that 'it's never enough.' People have jobs and families. It's OK if they join a vigil every once in a while. We tried to shift our understanding of what is a good activist. For me, it's someone who takes care of themselves, as well." This shift in understanding, she says, leads to more, not less, activism because it makes activism sustainable.

Activist Care

Bearing witness again and again can lead to secondary post-traumatic stress disorder (PTSD). In the early years, there was too little focus on self-care and community care, but now, as a movement, we are getting better. We are learning how to do things better and more sustainably. At the vigils in the Netherlands, Annette, a vegan psychologist, makes herself available to other activists, *pro bono*. She also holds workshops and creates videos on self-care. Lea says: "Previously, I took it upon myself to talk to everybody. So it's nice to have someone professional." As the country coordinator for the Netherlands, Lea travels the whole country to attend all the vigils, some of which are a few hours' drive away. Sometimes, she finds that she needs to take a step back. She is taking Greenpeace's community-based training on sustainable activism, the bottom line of which is: "Instead of doing as many actions as you can, it is taking care of activists that matters first and foremost. Activists are happier if we take care of each other, and this is more sustainable as well as more effective."

A lot of activists in our movement experience periods of anger. Lea observes: "It's OK to be angry sometimes, but anger shouldn't be our baseline. Our baseline should be love." Lea's tips include taking breaks from social media. Every two weeks, she puts her phone on airplane mode for two days. For Lea, it's important to take care of her body, work out, and do yoga. She also emphasizes a proper balance between activism and rest. "We need to make time to do other things. Choose to do something fun together instead of a vigil or square. It's never a wasted opportunity when you get rest and have fun. Go out for good food and drinks. The next day, you'd want to come back to activism. With social media, it's important to also follow happy accounts like those of animal sanctuaries."

Valezca Munsuri, an organizer for Mexico City Animal Save and Animal Save Movement's director of recruitment, struggled to find balance to make her activism sustainable. "When I started attending vigils, I was very focused. When I got home, I was really sad, but I didn't know how to deal with it." She says her past experiences with family had prepared her. "I understand what I need to do to keep going. And of course, I made a promise to the animals to keep working for them. *I realized that I needed to compartmentalize my life*. I will do activism, then I will go and rest, have a good meal, go out with my friends, and do what makes me happy without focusing on all the suffering." When going out, you don't want to end up centering conversations on negative things either. "Just talking about suffering animals is not healthy. We go to museums or have parties, and this gives me an escape. I am also always exercising and focusing on my wellness." This allows Valezca to continue to organize weekly animal vigils and to do so in a sustainable way. "It's still very sad, but I learned over time how to deal with this and make myself feel better and understand when to do activism and when to do other things."

When the Plant Based Treaty campaign was launched, Valezca embraced it as a solution-oriented initiative that complements her difficult work bearing witness at slaughterhouses in marginalized, poverty- and violence-stricken communities on the outskirts of Mexico City. "I still think it's important to show what happens at slaughterhouses and give the animals water, but on a daily basis, I like to work on plant-based diet solutions. I don't go crazy thinking about how the animals are suffering 24/7 and instead work on other campaigns." The various chapters in Mexico City (Animal, Climate, and Health Saves) organize direct actions, vegan food giveaways, and Plant Based Treaty city-endorsement initiatives, intermittently keeping activists' minds away from the slaughter.

Hoshimi Sakai, Animal Save Movement's incredible animator and a dedicated campaign organizer, sees a therapeutic quality in the arts. "Visual art, dancing, music, writing, poetry, or any other form of self-expression can be very therapeutic for a lot of people, though not a replacement for therapy." Hoshimi says many people see the arts as something extra or merely as entertainment. "It's more than that. People need food, water, and shelter to survive. But they also need to have hope and a sense of belonging within the movement to come together. Art is a part of activism." She challenges the idea that one is not a real activist if they don't go to protests. "Activism is like a symphony orchestra: everyone plays a role. The concertmaster can't play a symphony by themselves. You need the person in the back playing the triangle. It's a collective movement. People have different abilities. We just need everyone to recognize that everyone can do something."

Adrienne George, our Human Resources manager, has been organizing wellness workshops with a variety of psychologists and wellness coaches who are keen to help Animal Save Movement activists cope with all the suffering they witness. Wanting to develop a fuller program, she is working with Michelle Labrosse, a wellness coach and yoga instructor, who is perfect for the job. Michelle offers four yoga and meditation classes each week for all Save organizers as a way to enhance self-care and promote sustainable activism. Personally, I've already noticed a huge impact on my daily life. I find her classes have a calming and grounding effect on my mind and body and help restore my energy. It's also good to know that Michelle does the same activism as we do and her mission is the same as ours. Valezca adds: "I always knew we needed to move our bodies but had never met an expert in the field. Michelle is a very positive and vivacious person. Her classes include full-body stretching, combined with core strength exercises and meditation. It's a really complete exercise, and she's always saying positive things. If you say, 'I'm so lazy,' your consciousness then believes that. If you say positive things like, 'I can do this,' you will move forward."

Michelle Labrosse says sustainable activism requires two things: first, giving people a sense of belonging, and second, putting tools and resources in place so you can take care of yourself and others holistically. Burnout—when activists feel unable to continue due to mental and physical exhaustion—is less likely with kind, shared leadership; a love-based approach to activism; and friendships that emerge from actions and community events. Every new person at a vigil or another event needs to be greeted and introduced so they can feel they're part of the community. Also, activists suffering from PTSD need to be equipped with the tools necessary to develop coping mechanisms, which may include meditation, physical exercise, breath work, healthy eating and adequate hydration, among others. Michelle's yoga, exercise, and meditation sessions serve as a safe space for like-minded people with similar goals. Her aim is for activists to become people who lead by example, who take care of their mind, body, and soul. Her "Work out, Meditate, Activate" classes comprise a forty-five–minute workout and fifteen-minute meditation.

Azul Cardoso, an organizer for Montevideo Horse Save in Uruguay and another of our recruitment directors, feels that we sometimes push ourselves too hard when we bear witness to the suffering of animals. It's really important, she says, especially for people who have been activists for many years, to *develop a balance in life*. She agrees with Valezca in that one has to "have some fun and do other things with other communities." She copes with stress by trying to understand and work with others outside of the vegan bubble. Her chapter started a vegan soup kitchen that makes a hundred homemade burgers every Saturday for those in need. When you work with others and connect communities, according to Azul, "you know better how to live and spread the message of systemic change." Azul strongly believes that in her native Latin America, "animal rights groups need to be a part of the resistance against social inequality," as "an envisioned society that doesn't oppress animals can't be one that oppresses human beings."

Camille Savage, director of animal rescue for Animal Save Movement and organizer for Las Vegas Animal Save, was holding, without fail, two vigils every month in 2018–2019. The group was able to survive the challenges posed by COVID-19 because their activities have evolved. First, they had to adjust the vigils, then decrease the number of vigils. In late 2019, they started holding volunteer events at local animal sanctuaries once or twice per month. Then they started organizing cleanups at the local park after vigils to decompress. For Camille, the park cleanups are a way to connect with all the activists so they don't go home feeling powerless. In 2020, they downsized further and started holding vigils only once a month.

Camille says: "In Las Vegas, there are not a lot of people organizing for animals, yet there is so much need in a city that's an entertainment destination. There are a lot of vegan restaurants, but few street actions. Comfort breeds complacency. Being uncomfortable in places of exploitation is important." Camille does so much, but how does she cope? She articulates how many of us feel about sustainable activism when she says: "It's about effectiveness and survival. We have to check, first of all, if what we are doing is effective and, secondly, if we need more breaks. It's about knowing yourself and being OK with changing things up because you have to." Camille has adjusted vigil activities such that those who need space may focus on chalking and holding signs from a distance, while others stay on the sidewalk to watch and bear witness to the "processing" of chickens, document what they see, try to rescue any chickens who escape, as well as catch any infringements to report to government authorities. She is constantly experimenting with different events at animal sanctuaries, such as tree-planting and fundraising events. At the same time, we have to evaluate our own limits and boundaries. "The biggest thing about sustainable activism is balance," Camille says. This much-needed balance allows us to keep our movement animal-centric and to help more and more animals, while taking care of each other and building a vibrant and healthy activist community.

7

Going Vegan for the Animals, the Climate, and Your Health

There are so many reasons to go vegan—it was important to us that all of them are meaningfully reflected in Animal Save Movement's work. We embarked on a restructuring exercise to rebrand the organization and create new branches. In the spring of 2018, Federico Callegari, an organizer with Buenos Aires Animal Save and our graphic design coordinator, proposed renaming the organization from The Save Movement to Animal Save Movement. Alongside Animal Save Movement, whose mission is to hold vigils at every slaughterhouse in the world to help end the animal exploitation industry, we created Climate Save Movement, which aims to solve the climate crisis by ending animal agriculture, reforesting the Earth, and phasing out fossil fuels, as well as Health Save Movement, which promotes veganism as a way to prevent and reverse diseases related to animal product consumption and works to advance food justice. These intersectional campaigns make it easier to form broad-based coalitions and cross-cutting alliances working towards animal liberation and social justice.

Health Save Movement

Health Save Movement first emerged in Argentina in 2018. Mariana, our video director, remembers: "Returning from a vigil, we thought about cooking vegan food for people on the street. We were criticized for not helping people in our activism, so we started cooking in a kitchen, making twenty meals at a time with lentils and vegetables, and we opened Amigues por las calles, which means 'friends in the streets.' We spent a year doing that, then Nicolas Fassi, a medical student, helped set up Health Save Movement networks." In April 2019, the first Health Save chapter began in Caballito in Buenos Aires—a group of activists going out every Monday and handing out nutritionally complete meals. Vegan brands and businesses were contacted for food donations.

Health Save Argentina recruited vegan influencers to increase the number of volunteers and set up new chapters. Connie Isla, a vegan singer, actress, and activist, was the first social media influencer to participate in the vegan food giveaways and help spread the word. A few months later, several more chapters had been started in Buenos Aires and Rosario, and the number of volunteers had multiplied. Mariana says: "Connie helped with cooking and sharing our activism on social media. In two weeks, we grew from 0 followers to 10,000! Nowadays, she still participates. Whenever she comes, she films videos and helps us grow our social media networks! She is the biggest influencer who comes to do activism with us, but then we also have influencers who help us share content."

Mariana continues: "We are vegans fighting for animals and humans. We can be interested in both causes. I participate in weekly food giveaways, and I also help other networks. I collaborate with Health Save Argentina on social media. I also join Voicot [an art movement for animal rights; its name translates to 'boycott'] when they go out to paste up posters in the streets. Luckily, there is a lot of activism in the capital city, so there are always opportunities to participate. I help mainly with filming and editing, which is what people need from me in their activism."

Jimena Zamora, global director of Health Save Movement and lead organizer of Health Save Argentina, was born in Buenos Aires in the early 1980s. She always empathized with animals and became vegetarian when she was twenty after learning about the abuse in animal agriculture. She studied filmmaking at the University of Buenos Aires and worked for more than fifteen years in audiovisual production, mainly in advertising. When she became vegan in 2017, she started devoting her life to activism. "I started with Health Save to get involved with social justice and give animal liberation a fuller meaning. The two go together. There is no such thing as animal liberation if there is no justice for humans." She says: "Argentina is a very large country with many contrasts—with beautiful places and many prosperous people, but also a large population of people in vulnerable situations. The fact that there are people sleeping on the streets is partly related to agriculture and Argentina's being one of the countries that consume the most meat."

Jimena oversaw the growth of Health Save Argentina into a dozen chapters, with an average of ten activists at each chapter, which hold weekly activities. These chapters assist soup kitchens, community gardens, and farmed-animal sanctuaries, among others. One of Jimena's main tasks is organizing people's participation and ensuring they have all the tools to support their group. "Doing a weekly event is a lot of work; it would not be possible if there are no volunteers willing to join and no coordinators to support them." She says that they have a great community and that many companies are motivated to donate food items after seeing the success of their programs every day on social media.

Health Save Argentina campaigners deliver nutritious meals to those in need. Jimena reports: "We give away 500 vegan meals a week, taking into account all the chapters in Argentina. Most of them are in Buenos Aires. The two largest chapters are Caballito in Buenos Aires and Rosario in Santa Fe. Between them, these two distribute almost half of the meals." The food-giveaway program relies on the help of several donors who contribute ingredients and "all the volunteers who cook week after week." Care is taken to make sure that each dish served is nutritionally balanced, with cereals, legumes, and vegetables. Health Save Argentina expanded their outreach program by distributing bags of organic fruits and vegetables cultivated by farming families. They also organize food-sampling sessions in parks around Buenos Aires so that people can taste vegan foods and learn about the health benefits of a vegan diet. They have teamed up with Climate Save Argentina for a monthly Vegan Fair, a well-attended, one hundred percent vegan festival in Buenos Aires where they promote the Plant Based Treaty and give away native plants and seeds.

In addition, Health Save Argentina has built two organic community gardens outside soup kitchens in 2021 and is expanding the program with more gardens and compost bins set up in the Buenos Aires area. They are partnering with the soup kitchens on a project to plant native trees within the city. "It is a project whose results will be seen in the long term, but we are very excited about it." They are also piloting a "food sovereignty school," with classes taking place biweekly on Saturdays, where volunteers meet to learn about food sovereignty and agroecology. "It's a space for volunteers to get to know each other, learn, and later take that practical knowledge to other projects we have."

Jimena adds optimistically: "The movement continues to be sustained thanks to perseverance, the maintenance of a safe space, and the motivation to grow and add activities. During quarantine, we established alliances with soup kitchens and they very kindly accepted our vegan dishes. Eventually, we were able to start with the installation of vegetable gardens to nurture food sovereignty. We also organized talks with nutritionists and shared informative material."

A New Organizational Structure

Our movement stresses local, democratic power-sharing principles rather than a hierarchical, top-down structure. A formative concept since our first chapter has been expanding team leadership. The idea came from Cesar Chavez's approach to leadership and organization in founding United Farm Workers, which proved essential in the organization's growth and its ability to achieve many victories. Marshall Ganz, former campaign director at United Farm Workers, then a professor of sociology at Harvard University, writes that a large, democratic, representative leadership team that is open to insiders and outsiders (like traveling activists) alike is more likely to be innovative and to make effective use of a host of tactics.[44] Expanding leadership teams facilitates the rise of a global movement by welcoming new members as potential leaders and fostering leadership in everyone.

We were also influenced by Tolstoy's nonviolent anarchism, defined by political and economic democracy. We created a flat, horizontal organizational structure globally. Dozens of action and hub grants provide support for local leadership development in active cities worldwide. We set up a system of country and regional liaisons in North and South America, Asia, the Middle East, Africa, Europe, and Oceania to facilitate communication and the sharing of resources between local chapter organizers and local teams. This also lends itself to expansion of and innovation within the movement.

Every social justice movement in history has succeeded when enough people decided to stand up. To witness suffering animals and have them look you in the eye changes everything. By bearing witness to egregious injustice against animals and helping to create a new cultural norm whereby people see it as their duty to not look away, we discover the unity of life. By combining bearing witness with love-based community organizing, we quicken the pace of progress towards social, Earth, and animal justice. What started as a local organization focusing largely on bearing witness at animal vigils evolved into a movement that makes use of other, innovative actions: human chains or hugs around transport trucks, all-day and five-day vigils, hunger strikes, corporate research and the development of pressure campaigns with home demonstrations targeting slaughterhouse owners, and disruptions of shareholder meetings.

In order to continue to expand and improve our effectiveness, we realized there was a need to implement some significant structural changes within our organization. We set up a functional board of directors and added another layer to our structure by creating a representative, global core team of experts to coordinate campaigns for our various branches,

---

44 Marshall Ganz, *Why David Sometimes Wins: Leadership, Organization, and Strategy in the California Farmer Worker Movement*. Oxford: Oxford University Press, 2009.

animal rescues, regional vegan outreach efforts, communications, digital media, graphics, recruitment, and marketing. With a few changes to our existing horizontal structure, we saw huge improvements in our capacity to launch and deliver global campaigns such as Don't Look Away (encouraging the public to watch *Dominion*, *Seaspiracy*, and other life-changing documentaries), Stop Animal Gifting (demanding that development charities stop gifting goats, pigs, and chickens to impoverished areas of the world and instead gift plant-based food solutions), and most ambitiously, a Plant Based Treaty, as described in the following chapter.

After we opened our climate- and public health-focused branches and established our new, globally representative team of experts, we strove to develop Animal Save Movement's capacity to further move beyond the single tactical focus on vigils. We feel we are now ready to engage in significant and necessary campaigns to create mass mobilization, reach critical mass, and work towards gaining majority public opinion for animal justice.

8

Plant Based Treaty

To say "no" to animal exploitation is one reason to go vegan. To save the planet is another. You don't need to be a climate scientist to know that there is a climate crisis happening now. The past decade has seen record-breaking storms, forest fires, droughts, heat waves, and floods around the world—with just 1.1 degrees Celsius of global warming. We are heading towards climate chaos, spread of disease vectors, breakdown of ecological and agricultural systems, rising sea levels, and more frequent and severe extreme weather events such as hurricanes. These changes have even made it difficult to find good locations for new animal sanctuaries, as increasingly, there will be areas of the world that are too hot for humans and other animals to live in.

As hurricanes and other extreme weather events strike areas with industrial farms, the animals—pigs, cows, chickens, turkeys, and so on—are the forgotten ones. For example, in 2018, Hurricane Florence killed 3.4 million chickens and turkeys and 5,500 pigs in North Carolina, according to the North Carolina Department of Agriculture and Consumer Services. "The North Carolina Department of Environmental Quality said it had received reports of breaches, or structural failures, at at least two hog-waste lagoons. One breach in Duplin County was considered a total loss and more than 2.2 million gallons had spilled out. . . . Heavy rains had caused manure to spill over at twenty-one additional lagoons. . . .

Spilled waste from lagoons risks contaminating groundwater, including pathogens like salmonella, insecticides, and pharmaceuticals."[45]

In 2018, the United Nations Intergovernmental Panel on Climate Change (IPCC) stated that we had twelve years left to fix this global catastrophe. This means that we must take bold action in the 2020s to stop runaway climate chaos.

Animal agriculture is an enormous contributor to the current climate emergency. In August 2021, Animal Save Movement launched the Plant Based Treaty as an arm's-length, autonomous initiative that advocates individual and systemic change, with the aim of helping alleviate the climate crisis using plant-based food solutions.

In February 2022, the IPCC's second working group on climate impacts released a much-anticipated, three-thousand–page report that states dietary changes are an important solution to the current climate crisis and suggests policy measures accordingly. "People that eat high amounts of meat or unhealthy foods could reduce consumption of these foods and have more diverse diets. These dietary changes will benefit their health and reduce pressure on land. Regulated labeling, education and other policies which encourage healthy diets can support these shifts."[46]

Rising temperatures are already impacting food production, and there are concerns that further increases will lead to more crop failures. The IPCC's climate impacts working group prepared a detailed "Summary for Policymakers," which predicts, with high confidence, changes to ecosystem structures including freshwater systems, oceans, and fisheries. There are already observed and anticipated negative impacts on animal agriculture, crop production, and water supply in every region of the Earth. Beyond our food systems, there are negative climate impacts worldwide in terms of public health, including the increased prevalence

45 Kris Maher and Ben Kesling, "Florence Flooding Hits North Carolina Hog Farms Hard," *The Wall Street Journal*, September 19, 2018. https://www.wsj.com/articles/florence-flooding-hits-north-carolina-hog-farms-hard-1537398585. For more on the health impacts of bacterial infections, see https://www.peta.org/issues/animals-used-for-food/health-risks-meat-industry/

46 IPCC AR6 WG2, *Climate Change 2022: Impacts, Adaptation and Vulnerability*, Chapter 5, February 2022.

of infectious diseases, heat-related illnesses, and malnutrition. There are projected damages, including from inland flooding and storms, to cities, settlements, and overall infrastructure.[47]

There is a major greenwashing exercise underway in the animal agriculture industry, often drawing on the tactics of Big Oil. Industry proponents talk of methane masks to protect against cow burps, changes to animal feeds, deforestation pledges (which have since been backtracked on), and improvements to production. However, the IPCC impacts working group has assessed the feasibility of a sustainable "livestock system," among other possible climate adaptations, and has stated with a medium to high confidence that this measure has low feasibility.

Few of those in power are talking about animal agriculture in relation to climate. Yet, the Food and Agricultural Organization estimates that the industry contributes 14.5 percent of the globe's greenhouse gases. That's more emissions of noxious elements, such as carbon dioxide, methane, and nitrous oxide, than are caused by all cars, trucks, planes, ships, and other transportation modes combined. In 2019, the IPCC reported much higher emissions of carbon dioxide, methane, and nitrous oxide from animal agriculture, accounting for 21–37 percent of total anthropogenic greenhouse emissions.[48] In addition, there is the issue of opportunity costs from the loss of forests due to the meat, dairy, and egg industries. Today, almost half of all ice-free land is used for animal agriculture, grazing, and feed crops for the eighty billion land animals raised and slaughtered for food yearly. Forests absorb and store carbon dioxide, the primary greenhouse gas heating the planet. When we eliminate trees, we increase the amount of carbon dioxide in the Earth's atmosphere.

Prominent youth climate activist Greta Thunberg joined forces with Mercy for Animals to produce a public service announcement identifying animal agriculture as a leading cause of climate change and pushing for a shift to a plant-based food system to tackle this problem.

---

47 IPCC AR6 WG2, *Climate Change 2022: Impacts, Adaptation and Vulnerability: Summary for Policymakers*, February 2022, p. 11.

48 IPCC, Special Report on Climate Change and Land Use, Chapter 5, 2019.

We desperately need a *paradigm shift*. According to scientific philosopher Thomas Kuhn, scientific revolutions and paradigm shifts occur when anomalies shatter a society's old belief system.[49] When we acknowledge the huge ecological footprint of animal agriculture, the implications for policy are immense. It's not just cars—it's hamburgers, too. Both fossil fuels and animal agriculture are devastating our planet. A truly effective climate solution requires an end to animal farming and industrial fishing, as well as the subsequent reforestation of land. This reforestation would produce an immediate reduction in greenhouse gas emissions, starting a reversal of climate change. The crucial task of all climate activists is to bring this inconvenient truth to the attention of governments, nonprofit organizations, private industry, and the media.

Taking Action with the Plant Based Treaty

We have developed a Plant Based Treaty, inspired by the Fossil Fuel Non-Proliferation Treaty created by Tzeporah Berman and others at Stand Earth. In the 1990s, I had participated in the Clayoquot Sound and Great Bear Rainforest old-growth forest campaigns, through which I first met Tzeporah Berman, who now chairs the Fossil Fuel Treaty. I reconnected with her when I, along my colleagues Nicola Harris, one of our communications directors, and Genelle Palacio-Butler of Youth Climate Save, met her in April 2021. Tzeporah recommended that we target governments in our efforts to put animal agriculture on the climate agenda. The Fossil Fuel Treaty collects endorsements from individuals, groups, and cities as a means of generating bottom-up pressure on governments to negotiate a global treaty. Their treaty has been endorsed by the Dalai Lama and over one hundred Nobel Laureates, as well as dozens of major cities, including Barcelona, Kolkata, Los Angeles, London, Paris, Toronto, and Sydney.

49 Thomas Kuhn, The Structure of Scientific Revolutions, Chicago: Chicago University Press,1962.

After the meeting, our heads were spinning, and our team was inspired to soon begin preparations for the whirlwind soft launch of the Plant Based Treaty initiative four months later.

Upon the release of the IPCC's sixth scientific assessment report in August 2021, UN Secretary General António Guterres declared a "code red for humanity." There has never been a more urgent need for the Plant Based Treaty. The Paris Climate Agreement is largely silent on animal agriculture, despite the industry being a critical contributor to global warming and its eradication being an essential part of meeting climate goals. As the independent policy institute Chatham House observed, "shifting global demand for meat and dairy produce is central to achieving climate goals."[50] This position is strongly supported by scientists around the world, including Oxford University researcher Michael Clark, who points out that "even if fossil fuel emissions stop immediately, emissions from our food systems alone could increase global temperatures by more than 1.5 degrees Celsius."[51]

Animal Save Movement calls the Plant Based Treaty a sister treaty to the Fossil Fuel Treaty. Like the latter, the Plant Based Treaty is science-based and solution-oriented; uses a bottom-up pressure campaign approach by collecting individual, group, business, and city endorsements; and provides resources for the implementation of climate solutions at the corporate and city levels. The Plant Based Treaty is different in that it calls for both individual dietary change and policy change. While the Fossil Fuel Treaty emphasizes the imperativeness of systemic change, it pays little attention to the power of individual consumers. We believe that to alleviate the climate crisis also requires individuals to play their part, since switching to a vegan diet is the most important action individuals can take to reduce their carbon footprint.

50 Rob Bailey, Antony Froggatt, and Laura Wesseley, *Livestock – Climate Change's Forgotten Sector: Global Public Opinion on Meat and Dairy Consumption*. Chatham House, December 2014.

51 Michael Clark et al., "Global Food System Emissions Could Preclude Achieving the 1.5° and 2°C Climate Change Targets," *Science*, November 6, 2020, pp. 705–8.

Research conducted by the Jump Campaign shows that both individual and systemic changes are critical to keeping within 1.5 degrees Celsius of warming. "Individual lifestyle changes can lead to a 25 percent reduction in global emissions, but the majority of the changes will come from systemic changes implemented by governments and the private sector."[52]

The Plant Based Treaty is working to create international policy change, focusing on phasing out animal agriculture and commercial fishing to fast-track progress towards a safer, plant-based food system. This will require unprecedented international cooperation in three main areas: ending the expansion of animal farming, ending government subsidies to animal agriculture, and reforesting the Earth.

Among the highlights of the Plant Based Treaty are the following—called the "3R"—demands:

Demand 1 ("Relinquish") focuses on halting the expansion of animal agriculture to stop the problem from getting worse. This means:

- No deforestation for animal agriculture;
- No building of new animal farms;
- No building of new slaughterhouses;
- No expansion or intensification of existing farms;
- No conversion of plant-based agriculture to animal agriculture;
- No conversion of any land for animal-feed production;
- No clearing of forests or other ecosystems for animal grazing;
- No new fish farms or expansion of existing farms;
- Protection of Indigenous peoples and their land, rights, and knowledge;
- A ban on all fishing in designated Marine Protected Areas;
- No new large-scale industrial fishing vessels.

Demand 2 ("Redirect") focuses on incentivizing plant-based foods by redirecting subsidies away from animal agriculture in favor of plant-based agriculture. Governments might also redirect financial support by using tax dollars and publicly funded public information campaigns to actively

52 Matthew Taylor, "Six Promises You Can Make to Help Reduce Carbon Emissions," *The Guardian*, March 7, 2022.

promote the transition from animal-based food systems to plant-based food systems.

Demand 3 ("Restore") focuses on rewilding forests, oceans, peatlands, wetlands, and mangroves to assist climate change mitigation. Actively fixing the problem and healing our planet helps build resilience against climate change. Proactive reforestation on a global scale, with the involvement of governments and international agencies, and retraining initiatives for farmers to transition away from animal agriculture will help lower atmospheric greenhouse gas concentrations.

Just like the Fossil Fuel Treaty, the Plant Based Treaty has a growth strategy that includes an arm's-length, white-label initiative. This means that any individual, group, business, or city can adopt the campaign and rally behind a single call to action demanding a global treaty. As the US animal protection organization In Defense of Animals declared, "We are all Plant Based Treaty."

We call upon civil society, businesses, and cities to endorse the global Plant Based Treaty, and national governments to negotiate and ratify it. On our website, we are aiming to collect endorsements from 1,000,000 individuals, 10,000 groups, 10,000 businesses, and 50 cities by 2023 in order to generate bottom-up pressure on governments to begin negotiations. In 2023, the UN is undertaking the very first global stocktake of the Paris Agreement, assessing our collective progress from 2021 to 2023, and the process will be repeated every five years. As such, in 2023, we will revisit and reset the Plant Based Treaty targets for the next five years.

By June 2022, 40,000 individuals, 800 organizations, 60 businesses, and 16 cities and villages had visited our website to endorse the Treaty. By building a globally representative and inclusive community of individuals, groups, businesses, and cities all standing behind the Plant Based Treaty, we can put pressure on governments to fight the climate crisis we are all facing.

Worldwide, our organizers distribute business cards at events with a QR code that takes people straight to our website to endorse the Treaty on the spot. Around 25–30 percent of the individual endorsers of the Plant Based Treaty declare their diet as omnivore, vegetarian, or pescatarian. They are sent a special email inviting them to take a vegan pledge for the planet.

Nilgün Engin, organizer for Ankara Animal Save and our director of vegan outreach for the Middle East and Northern Africa region, is developing innovative best practices for her team in Turkey to collect individual endorsements for the Plant Based Treaty. She says her team uses many methods, including reaching out to social media followers individually, holding documentary screenings, and hosting breakfast and dinner get-togethers. Nilgün has run a host of workshops on a variety of topics, from composting, making homemade plant-based milk and yogurt, to planting gardens, with some workshops geared specifically towards refugees. She asks workshop participants to endorse the Treaty. Her team members have also given presentations on the Plant Based Treaty at cafés and universities, and held in-person and online events with nonvegans. Nilgün says their approach to innovation is two-fold: first, they ask community members and followers to fill out questionnaires, then create new events based on their specified interests; second, they meet with team members across the country to discuss their progress. "We highlight the methods that have been going well and build on those actions. We try to diversify them and adapt them to other cities, or we implement them the same way if they work well. We also discuss obstacles. We try to come up with useful solutions to change the previous outcome. If a certain method proves to be absolutely dysfunctional, we consider not pushing it harder to avoid losing time and energy. Furthermore, any idea counts. We do not shut down any suggestions from our meetings. We try to come up with ways to make these ideas implementable."

Groups that endorse the Treaty include the Climate Emergency Institute, Oceanic Preservation Society, Fridays For Future Digital, ProVeg International, Viva!, A Well Fed World, PETA, and chapters of Greenpeace, Extinction Rebellion, and Fridays for Future. We are targeting more environmental groups, including seventy-five groups in the Friends of the Earth network. We hope to get support from additional chapters of Fridays for Future, Greenpeace, Sierra Club, and Extinction Rebellion to help secure their organizations' endorsements at the national and international levels.[53]

Our 600-plus business endorsers include Ecotricity, Forest Green Rovers football team, Heura, Linda McCartney Foods, Odd Burger, Tofurkey, VFC, Vegan Kind, and Veg Capital. Matthew Glover, co-founder of VFC, Veg Capital, and Veganuary, believes that businesses "can play a critical role in combating climate change," adding that "the Plant Based Treaty offers a logical pathway to a plant-based world." Sixty-two percent of consumers want companies to take a stand on the social, cultural, environmental, and political issues that they care about the most.[54]

Notably, we have endorsements from four Nobel Laureates, including Klaus Hasselmann, who received the Nobel Prize in Physics in 2021 for his studies on the physical modeling of Earth's climate, helping to quantify variability and reliably predict global warming; IPCC authors and expert reviewers; celebrities; doctors and healthcare professionals; elite athletes, including Olympic medalists; politicians from different parties across the political spectrum; and prominent faith leaders.

It didn't take long for cities to endorse the Plant Based Treaty. Ten days after the soft launch of the Treaty on August 31, 2021, the City of Boynton Beach in Palm Beach County, Florida, became the world's first city to give its endorsement. Rebecca Harvey, the city's first sustainability coordinator, said: "The City of Boynton Beach has committed to reducing community-wide greenhouse gas emissions by 50 percent by 2035 and achieving net zero by 2050. Promoting plant-based diets is an important way to help meet this goal and help improve the health and sustainability of our city." The City of Rosario endorsed the Plant Based Treaty at the

53 Nilang Gor, "Join the New Sustainable Food and Agriculture Committee," Sierra Club San Francisco Bay, February 15, 2022.

54 Afdhel Aziz, "Global Study Reveals Consumers Are Four to Six Times More Likely to Purchase, Protect and Champion Purpose-Driven Companies," *Forbes*, June 17, 2020.

Conference of the Youth, organized by Fridays for Future Argentina, in September 2021. By mid-2022, Aprajita Ashish, our regional coordinator for Asia, and her team had secured the endorsements of fourteen cities and villages in India.

Animal Save Movement's global team and worldwide network of chapters have been busy contributing to all phases of implementation of the Plant Based Treaty. Through collecting endorsements, we also recruit and mobilize our supporter base. Our newsletter gained 60,000 subscribers within the first year of the launch of the Treaty. Further, our growth strategy includes building alliances and creating Plant Based Treaty teams in different cities and countries.

An initial, pre-launch phase lasted four months from May to August 2021, shortly after we held a meeting with Stand Earth's Tzeporah Berman in late April. In preparation for the soft launch of the campaign, we developed our vision of the Plant Based Treaty through the "3R" demands—"Relinquish," "Redirect," and "Restore."

The launch of the Plant Based Treaty required the creation of its very own brand identity so that it could be adopted by individuals, groups, or businesses as their own. The Treaty was named after the Fossil Fuel Treaty model, which was developed in 2019, a couple of years before our initiative. We chose a name that would appeal to people regardless of diet and that we felt would give us better access to schools, board rooms, and city halls. Then, our branding expert, José Legato, worked to create a visual identity that represented a collective, community-led movement that advocates plant-based solutions to the climate crisis based on a serious, science-led approach. The design reflects the positive tone of our solution-focused campaign and takes inspiration from nature, deliberately using organic shapes and colors reminiscent of the natural world. The iconic "green thumbs up" uses the internationally recognized symbol of agreement to enhance the global reach of the campaign. This visual branding has been woven into many aspects of the campaign—from graphics, video content, to street actions.

Similar to the Fossil Fuel Treaty's website, our website has a campaign hub to assist and connect our global network of supporters, including groups and individuals. It's full of useful resources and actionable items, including guidance on meeting with local politicians, holding rallies, and employing digital media activism. Here, individuals, community groups, businesses, and cities can find the help that they need to implement plant-based climate solutions locally. There are sample letters, guides, and pitches to help campaigners collect Plant Based Treaty endorsements and launch city campaigns.

In addition, a team of doctors and nutritionists helped us put together two guides to help people transition to a vegan diet. There is a guide called "Nutrition for our Children," which gives parents all the information they need to raise their children vegan. There is a "Vegan Action Starter Kit," which is meant for those who want to become vegan but are not sure where to start. This kit is an easy, factsheet-style read that provides all the tools and nutritional information to get someone started. In development are new resources on food giveaways, community gardens, and vegan dog and cat food to expand our offerings.

Upon the soft launch of the Plant Based Treaty on August 31, 2021, we presented our new website and held global actions in more than a hundred cities, including Manchester, Bristol, Amsterdam, Lisbon, Rome, Los Angeles, Mexico City, Toronto, Seoul, Mumbai, and Ankara. Groups around the world organized rallies outside city halls to call for their cities' support, and politicians were invited to attend to learn more about the Plant Based Treaty. At a rally in Mexico City, Senator Jesusa Rodriguez said, "To know if someone really wants to fight for justice, you have to see what they eat."

COP26 in Glasgow

Since the UN Framework Convention on Climate Change (UNFCCC) was signed in 1994, world leaders have been meeting nearly every year at the Conference of the Parties (COP) to work on a global response to the climate emergency. In 2021, we decided to travel to COP26, hosted in Glasgow, Scotland, and introduce the Plant Based Treaty there. We began by joining and tabling at the Conference of the Youth at Strathclyde University; then, we attended COP26 (October 31–November 12) to raise awareness of the impact of animal agriculture and the need for something like the Plant Based Treaty. The Glasgow Plant Based Treaty team received logistical support as well as support from a remote team of communications and digital media experts. Jens Holme, a Swedish MP, said, "The launch of the Plant Based Treaty was one of the most promising events at COP26."

We were present both inside and outside COP26 to ensure the impact of animal agriculture on climate was not ignored.

We created a Plant Based Treaty guide to Glasgow and launched it on November 1 to coincide with World Vegan Day and the start of negotiations at COP26. Thirty thousand copies were distributed within COP26, urging delegates to eat plant-based foods during COP26 and beyond, as well as around Glasgow.

Early during our stay in Glasgow, we held a pig vigil at Ardrossan slaughterhouse. More than a dozen activists stood outside the slaughterhouse holding Plant Based Treaty signs and banners. It was Lia Philips's first time bearing witness. "We couldn't believe what we were witnessing. We stood filming for five minutes. I recorded a video of two pigs comforting each other, nuzzling and pawing like two puppies. The driver of the truck then asked us to step away and let him drive into the slaughterhouse grounds. Helplessly, we stepped back. Weeping and hugging, knowing that we couldn't do anything to save them. After some time, we witnessed the clamber and screaming from the pigs as they were pulled from the truck and shoved into the building. I can still hear them screaming in fear, with the pure stench of death awaiting them—170 pigs in total." Those 170 pigs are now dead. Lia feels she should have done more to save the pigs she and other activists bore witness to, but she knows that is near impossible in today's society. "The only way to save their future ancestors," she says, "is to move towards a plant-based food system."

Our team in Glasgow carried out daily light shows downtown in George Square, where activists performed a Plant Based Treaty dance. We partnered with The Vegan Kind and Oatly to distribute thousands of free vegan food and drink samples to passersby. On the last few days of COP26, activists toured Glasgow with a five-meter inflatable cow to call on world leaders to stop ignoring "the cow in the room," attracting mass media attention.

To gain wider traction, we held two press conferences about the Plant Based Treaty in the Blue Zone at COP26—the first with Dale Vince, UN Ambassador and CEO of Ecotricity, and the second with ProVeg International, Oatly, and *Harry Potter* actor Evanna Lynch. Over one hundred interfaith leaders signed an open letter calling for global veganism and for COP's adoption of the Plant Based Treaty as a companion to the Paris Agreement, which received coverage by the *Jerusalem Post.* Stephanie Cabovianco, our Climate Save campaigner, was part of a panel alongside ProVeg International, rugby league player Anthony Mullally, and food giant Upfield to talk about veganism. The panel session was hosted by adventure sports brand EXTREME and livestreamed to its 20 million followers.

Internationally-acclaimed recording artist Moby issued a video statement calling on world leaders at COP26 to negotiate the Plant Based Treaty, leading to follow-up interviews with the British *GB News* and the Australian *Times Radio* about the need for a plant-based food system to alleviate the climate crisis. The US organization In Defense of Animals helped us secure the endorsement of the McCartneys. Sir Paul, Mary, and Stella McCartney, along with Ela Gandhi (Mahatma Gandhi's granddaughter), issued a joint statement in support of our campaign, which attracted widespread mass media coverage, including by *The Guardian* and on musical platforms like *NME* and *Billboard.* Even conservative media outlets, such as *The Times*, wrote about the McCartneys, reaching audiences not often exposed to news stories highlighting plant-based climate solutions. Overall, coverage of the Plant Based Treaty before and during COP26 (August–November 2021) across mainstream English-speaking publications had a reach of over 20 million.[55]

Bonn Climate Conference

Bonn is home to the headquarters of the United Nations Framework Convention on Climate Change (UNFCCC) and also where climate negotiations take place every year prior to the annual Conference of the Parties (COP). In 2022, about a dozen of our activists attended the Bonn Climate Conference to remind everyone at the UNFCCC that we need a plant-based treaty. Our first position paper, which we developed for the conference, was called "Appetite for a Plant-Based Treaty? The IPCC Repeatedly Demonstrates that a Vegan Diet Is the Optimal Diet to Drastically Reduce Food-Related Emissions" and was backed by over one hundred groups including Green Rev Institute, Future Food 4 Climate, and Oceanic Preservation Society.

We made three key points in our position paper. First, we showed that the IPCC repeatedly demonstrates that a vegan diet is the optimal diet for reducing greenhouse gas emissions related to food production, and that the Plant Based Treaty offers a roadmap for a fast and just transition to a plant-based food system in the 2020s in response to the climate emergency. For example, we pointed to the IPCC's 2020 *Special Report on Climate Change and Land Use*, which illustrates how we could save almost 8 gigatons' equivalent in direct greenhouse gas emissions each year if we all adopted a vegan diet.

Second, we called for more concerted efforts to properly address the methane emergency and reduce methane emissions, 32 percent of which are from animal agriculture. At COP26 in Glasgow, the US government initiated the Global Methane Pledge, which 112 countries signed, committing to cutting methane emissions by 30 percent by 2030. However, this still falls short of the 45 percent cut needed by 2030 to keep the global rise in temperature within 1.5 degrees Celsius, according to the UN. Furthermore, the states that have signed on to the pledge

55 Eden Green PR, *Plant Based Treaty Campaign PR Report*, November 2021.

are together responsible for only about half of human-caused methane emissions, while the pledge itself is voluntary and unenforceable, and it does not offer a roadmap to meeting the targets.

Third, we called for the transition to a plant-based food system to stop further deforestation and allow for massive reforestation to combat climate change and biodiversity loss. Since the Declaration on Forests and Land Use was signed by 141 countries in 2021, deforestation attributed to animal agriculture has continued unabated. Destruction of the Amazon in the first four months of 2022 hit record levels, with 1,954 square kilometers deforested, representing an increase of 69 percent compared to the same period in 2021. According to a famous study in Science by Joseph Poore of Oxford University and Thomas Nemecek of Agroscope in Switzerland, almost 80 percent of farmland is used to rear animals to produce just 18 percent of global calories. Clearly, it is hugely inefficient, wasteful, and polluting to produce or consume meat, dairy, and eggs. The production of beans, peas, and other plant-based proteins has a much lower impact and is far more beneficial for the environment.

For three days during the first week of the conference, in early June 2022, Animal Save Netherlands' bright orange food truck was parked right at the entrance of the UNFCCC building in Bonn. Next to the food truck was our inflatable cow with the messages STOP ANIMAL FARMING and ENDORSE THE PLANT BASED TREATY. Nonstop and with big smiles, the Plant Based Treaty team handed out an average of 600 free plant-based sausages to endless lines of delegates each day. Lea says: "People kept coming back and said we changed the atmosphere there. The fact that there was no plant-based food at the conference was a topic of discussion for a lot of people. On one day, there were only carrots and a salad. People were coming to the truck with praying hands, they were so relieved to have us there."

Meanwhile, our team inside the conference focused on getting delegates and attendees to visit our food truck. Fellow organizer Raphael at ProVeg International told Yael, our Plant Based Treaty coordinator, that they were really happy about the truck's presence. It helped conference attendees connect the plant-based solutions we were calling for to what was going on inside the conference walls. Yael says: “There was nothing, no vegan food inside the conference. Everyone was super happy that we had a food truck.” Only the cafeteria caterers were not so happy because sales were down! Lea says the complaints from the kitchen were good because this way, they would think about what to serve next year. She noticed there were people who hadn’t come to our truck on the first two days but had heard stories about how good the food was and decided to come on the third. Our team enjoyed handing out food more and more each day, and the appreciation kept growing. For Lea, the cherry on top was when a senior UNFCCC official came by for second helpings and said, “You are making a bigger impact than you think.”

Meanwhile, Yael says she was spending every minute of her days putting our Plant Based Treaty business cards and position papers on tables and in meeting areas. “The first day, the cleaning staff didn’t clear the flyers, so we just put the position papers out in every hall and in front of every chair in the conference rooms that we could find.”

Our team joined other groups in protesting the Koronivia Joint Work on Agriculture (KJWA) negotiations (the outcomes of which would be reported by the UNFCCC Subsidiary Body for Implementation, which oversees the implementation of the Paris Climate Accords), as some country delegates were trying to eliminate important sections from the negotiations, such as references to agroecology. Indeed, the KJWA’s agenda seems to be driven by animal agriculture. ProVeg International notes in its “2020 and Beyond” summary that the KJWA’s idea of sustainable agriculture and food security entails “improving livestock management systems, including agropastoral production systems in which the growing of crops and the raising of livestock are combined.” The possibility of transitioning to a plant-based food system is completely overlooked. Alarmed by the KJWA’s bias towards animal agriculture, we were there with our Plant Based Treaty banners and placards to call for plant-based agricultural solutions to be included and prioritized, and the media was there to record it.

At one point, while everyone was working hard, networking and speaking to national delegations and UN staff, Stephanie, our Climate Save campaigner from Argentina, and Yael noticed the press gathering around a man. After Stephanie confirmed it was COP26 president Alok Sharma, Yael said, "Let's go and give him a position paper." Despite wearing heels, she ran across the entire compound and cut through grass to hand our press release and position paper to the COP26 president, then to all the press present too for good measure. Not only that, Nilgün, our campaigner from Turkey, was able to talk to both the COP27 president and the executive secretary of the UNFCCC, persuading them to take our position paper as well. Yael says that she left her first climate conference very frustrated, with how slow policy-making processes are and how difficult it is to say anything. Nonetheless, she says, "We are on the right track in bringing people in, and it's super important to have young people lead the way."

9

Think Globally, Act Locally

LOCAL ACTIONS ARE KEY IF we are to curb the climate crisis. Waiting for negotiations among world leaders to succeed is not a wise approach. The Plant Based Treaty offers the "3R" solutions, which allow all societal actors to play a role by, for example: introducing pressure campaigns aimed at companies to stop the expansion of animal agriculture; launching city campaigns; targeting subsidies; working to veganize menus at schools and universities; running ad campaigns; organizing plant-based food and seed giveaways; and developing community gardens.

The first "R," *Relinquish*, aims to halt the widespread degradation of critical ecosystems caused by the expansion of animal agriculture. Based on the research and investigations we had conducted, we launched a number of digital media campaigns, including petitions, direct actions, and corporate campaigns, to stop animal gifting and octopus farming.

About 50,000 people signed a petition opposing Spanish multinational firm Nueva Pescanova's plans to open the world's first octopus farm in Gran Canaria. The 52,000-square-meter farm would lead to 1,000,000 octopuses being killed for food each year (3,000 tons, according to the company, with an average final weight of 3 kg per individual). Octopus farming is as unsustainable as it is exploitative. The proposed farm would put more pressure on the ocean and disrupt its ecosystems, not to mention directly killing countless marine creatures who would have to be caught and fed to the octopuses. To voice our opposition more strongly, in February 2022, Gran Canaria Animal Save and Climate Save held demonstrations outside the company's facilities.

Octopuses are the world's most intelligent invertebrates, who can make tools out of shells and build defensive rock structures outside their dens to protect themselves while they sleep; by some measures of intelligence, an octopus is as smart as a golden retriever. Dr. Jennifer Mather, a researcher on cephalopod intelligence, says octopuses have long-term and short-term memories and feel pain, perhaps even more extensively than do vertebrate animals. "They can anticipate a painful, difficult, stressful situation—they can remember it. There is absolutely no doubt that they feel pain. The octopus has a nervous system which is much more distributed than ours. If you look at us, most of our neurons are in our brain, and for the octopus, three-fifths of [his] neurons are in [his] arms."[56] The planned farm wouldn't allow the octopuses any natural areas, where they could forage for shells to build dens; if anything, it looks more like an industrial chicken factory farm. It's morally wrong to hurt any being who can feel pain. But especially in the case of octopuses, whose communities in the wild have been in decline, we should be helping to restore wild octopus populations rather than creating these prisons for breeding and exploiting them on a mass scale.

In 2021, we launched Stop Animal Gifting, targeting development aid charities such as Oxfam, World Vision, and Heifer International with an online campaign and direct actions outside their offices in multiple countries. Several well-meaning organizations promote sending live farmed animals as "gifts" to low-income communities and regions, with the intention of helping them reduce hunger and poverty. Many of these programs are huge, with Heifer International alone gifting over 720,000 animals in 2020. Sending goats, cows, chickens, and other farmed animals as gifts may seem like a simple act, but its true cost includes environmental degradation, soil acidification, water contamination, air

56 Eric March, "Octopus Intelligence: Here Are 13 of the Most Frighteningly Smart Things They Can Do," *Upworthy*, April 20, 2016; Hilary Pollack, "How an Octopus Feels When It's Eaten Alive," *Vice*, December 2014.

pollution, deforestation for animal feed, forest fires, a worsening climate crisis, bird flu outbreaks, human health problems such as diabetes and heart disease, the psychological toll of more community slaughterhouses, and even childhood trauma in the case of children who have to watch their beloved animals get brutally slaughtered. The Stop Animal Gifting campaign aims to encourage individuals to instead give compassionate, plant-based gifts, and development aid organizations and charities to implement plant-based projects to both alleviate poverty and tackle the climate, ocean, and biodiversity crisis.

In the spring of 2021, Animal Save Movement India campaigners Aprajita and Luckysh visited the state of Odisha, where Cargill had partnered with Heifer International to convert predominantly plant-based tribal communities to chicken farmers. In late 2021, we began reaching out to the heads of development aid charities to express our concerns about their animal-gifting programs and arrange meetings with them to discuss alternatives. These charities can alleviate poverty and hunger by creating community seed hubs, installing water irrigation systems, providing training on permaculture and veganic farming techniques, reforesting, regenerating the soil, planting trees to increase canopy to help improve the water cycle, restoring savannah to rainforest, and restoring other key ecosystems. All contacted charities declined requests to meet with Animal Save Movement ahead of our campaign launch to discuss the rollout of plant-based aid programs, which would allow recipients to grow crops for direct consumption instead of as animal feed, thus providing more food and more food security.

We then launched our direct action and online campaigns in the run-up to the holiday season, urging charities to heed our call and switch to plant-based development aid. We held protests outside Oxfam and World Vision offices in Ottawa, Toronto, and Mumbai.

We developed a series of video testimonials from former donors who expressed their disappointment with the charities' continued contribution to animal cruelty and the climate crisis, as well as from activists in recipient countries like India, Mexico, and Kenya. Stephanie

Schwartz said: "For many years, my husband and I sponsored children from around the world through an organization called Plan International Canada. We wanted to help children with healthcare, food, water, education, and vegetable gardening. . . . It seemed to help the community. We even got updates from the kids themselves; some of them were artists. One child we especially bonded with because she loved animals, and we argued whether we liked dogs or cats better. Then, one day, we heard from the organization that they were going to start animal gifting. It got me really worried because I was picturing an animal tied up in the backyard, an animal drinking the family's water and eating their food. It's not sustainable. They had been doing vegetable gardening before that. And besides that, it's violence and sacrifice, and the kids would be screaming when they saw the animal slaughtered. I would never give an animal like this. . . . We wrote to Plan International Canada; they wouldn't stop, so we had to stop."

Rachel Kabue, organizer for Animal Save Kenya and Animal Save Movement's regional liaison for Africa, commented in a video statement: "I'm concerned about the climate crisis here in Kenya, where we face rising temperatures and the worst shortage in rainfall recorded in decades, which has wiped out pastures and caused food shortages. . . . I'm asking Heifer International to stop animal gifting. . . . The spread of animal agriculture is contributing to the climate crisis in Africa."

We developed action alerts, email letters, and a petition calling on Oxfam, World Vision, Heifer International and Cargill's Hatching Hope project, Christian Aid, Save the Children, Plan Canada, Lutheran World Relief, Canadian Feed the Children, Tearfund, and other development aid charities to commit to carbon disclosure in all of their projects, stop animal gifting, and implement projects promoting a plant-based food system as crucial steps in addressing the escalating climate crisis. Thousands engaged with our campaign page, StopAnimalGifting.org, signing our petition and sending email letters.

In addition, Animal Save Movement teamed up with In Defense of Animals' (IDA) Interfaith Vegan Coalition to put together open letters from scientists and interfaith leaders. Number one on our list of spokespersons was Dr. Jane Goodall, whose participation IDA's Lisa Levinson helped us secure. She and Dr. Goodall had worked on a wetlands protection initiative together in Los Angeles.

Along with fellow scientists and religious leaders the world over, Dr. Jane Goodall, DBE, founder of the Jane Goodall Institute and a UN Messenger of Peace, urged international aid charities to end animal gifting. Dr. Goodall filmed a video statement in which she raised concerns about the unintended consequences of gifting animals to poverty-stricken people in the Global South: "In the lead-up to Christmas, many people are feeling generous and want to help those less fortunate than themselves. There are a number of organizations that have launched campaigns, suggesting that one way to help those suffering poverty and hunger is to gift them an animal, such as a heifer. As a result, farm animals are purchased in great numbers by generous donors. Unfortunately, this can result in unintended consequences. The animals must be fed and they need a lot of water, and in so many places water is getting more and more scarce thanks to climate change. Veterinary care is often limited or totally lacking. It will be ever so much better to help by supporting plant-based projects, and sustainable irrigation methods, regenerative agriculture to improve the soil. Well, this means charities must develop plans to create a gift package that will appeal to the generosity of those who want to help those less fortunate than themselves. Thank you."

Dr. Goodall's message to development charities worldwide received coverage in about thirty publications including *The Guardian*, *Dorset View* (in Dr. Goodall's hometown), *Church Times*, *The Beet*, *Plant Based News*, *Green Queen*, and *One Green Planet*.

The second "R" (*Redirect*) aims to promote a universal shift to healthier, more sustainable, plant-based diets. We organize city campaigns and ad campaigns promoting plant-based eating, target food subsidies, and work to veganize menus at schools and universities.

Cities face significant threats as a result of climate chaos, including threats of heat waves, extreme weather events, and rising sea levels (for most coastal cities). City governments also significantly impact greenhouse gas emissions, public health, and animal protection, having jurisdiction over land use, housing, food, and, in general, production and consumption activities. In the Netherlands, for example, local authorities have a sphere of influence that extends to 40 percent of greenhouse gas-emitting activities.[57] In the US, more than 470 city mayors make up the group Climate Mayors, which helps "build political will for federal and global climate action" and promotes best practices among cities. These mayors have agreed to develop a community greenhouse-gas inventory, set short- and long-term targets to reduce emissions, and develop a climate action plan aligned with their city's targets.

Globally, about one hundred cities have signed up to become part of the "C40 cities" and to "take urgent action to confront the climate crisis." The C40 website states that its mandate includes helping "cities to implement solutions that make it easier for people to eat more plant-based options and waste less food." Several C40 cities have committed to achieving the EAT-Lancet Commission's Planetary Health Diet for their citizens by 2030. The diet recommended by the EAT-Lancet Commission on Food, Planet, Health is predominantly plant-based, with half the plate consisting of fruits and vegetables and the plant protein portion being three times bigger than the amount of animal protein. EAT-Lancet's home page says this is the optimal diet—but not quite! The optimal diet is vegan, according to the IPCC's 2019 *Special Report on Climate Change and Land*. A global shift to a vegan diet could reduce food-related greenhouse gas emissions by almost 8 gigatons of carbon dioxide equivalent a year, whereas a shift to a flexitarian diet of limited meat and dairy would reduce emissions by only 5 gigatons. Further, a 2020 study found a vegan diet is optimal and significantly better than even the EAT-Lancet Commission's mostly plant-based diet, offering a "double carbon dividend" from both

57 Frans Coenen and Marijke Menkveld, in *Global Warming and Social Innovation: The Challenge of a Climate-Neutral Society*, ed. Marcel Kok et al. London: Earthscan, 2002, p. 123.

lower emissions from food production and increased carbon sequestration by freed-up land. The study found that "shifts in global food production to plant-based diets by 2050 could lead to sequestration of 332–547 gigatons of carbon dioxide, equivalent to 99–163 percent of the carbon dioxide emissions budget consistent with a 66 percent chance of limiting warming to 1.5 degrees Celsius." As part of the Plant Based Treaty campaign, we are trying to convince city and national governments to aim for a vegan diet—the truly optimal diet—for all citizens.[58]

We are simultaneously running city endorsement campaigns to ultimately pressure national governments to negotiate a global plant-based treaty, and trying to influence city policies and promote plant-based solutions at the local level. Our recruited local teams are helping us secure pledges from fifty cities by 2023, influence cities' climate action plans, shift city expenditures away from animal products and towards more plant-based products, investigate whether city council pensions are invested in animal agriculture, and launch community education campaigns. We have launched multiple campaigns in the UK—in Glasgow, Bristol, Haywards Heath, and Manchester—hoping to persuade these cities' authorities to table a motion endorsing the Plant Based Treaty.

India's Plant Based Treaty cities

In India, fourteen cities and villages in four different states have endorsed the Plant Based Treaty. Aprajita says her strategy was to get one city to endorse the Treaty, then ask the mayor for a reference to get other cities to give endorsement in a domino effect. She also focused mostly on cities in the state of Gujarat, a heavily vegetarian state with a Jain majority (Jainism is a religion that follows ahimsa practices including towards animals). The fourteen cities that have signed on to our Treaty include nine cities in Gujarat (Ahmedabad, Bhavagar, Bhuj, the village Bhujpur, Gandhinagar, Jamnagar, Mundra, Rajkot, and Vadodara); Amravati, Nagpur, and Thane

58 IPCC, Figure 5.12; Matthew N. Hayek, H. Harwatt, W. Ripple, and N. Mueller, "The Carbon Opportunity Cost of Animal-Sourced Food Production on Land," *Nature Sustainability*, September 7, 2020.

in the state of Maharashtra, where Mumbai is located; and Jabalpur in Madhya Pradesh and Sonipat in Haryana, both in central India.

Aprajita and her team are being asked by city mayors to assist in educating the public. They are also being offered platforms in government circles and in schools and universities. Aprajita found that many government officials were aware of neither the relationship between animal agriculture and the climate crisis, nor the sheer amount of freshwater used by this industry. Aprajita says: "We use 70 percent of freshwater for animal agriculture. They say everywhere that we need to save water. It's in their comfort zone to focus on individual use." One cabinet minister from Rajasthan contacted Aprajita after hearing about our efforts. He told her: "We have lots of schools and colleges under us. If you want to be a speaker there, we would be very happy." Aprajita and her team are being introduced to education departments at universities and invited to run educational workshops on climate and plant-based climate solutions; they are asked to paint murals at schools. A mayor in Gujarat asked for the team's help with all fifty-five schools in the city. The opportunities are abundant and it's getting difficult to keep up, so Aprajita looks to expand her team.

During her third visit to Treaty-supportive cities, Aprajita set up teams in three cities. She was able to get in touch with many organizers through the National Animal Rights Day (NARD) events, held annually in early June. NARD has chapters in nineteen cities in India, though its focus is street action. Aprajita says: "Before, these activists were doing street outreach; now, they are excited to paint school murals with Plant Based Treaty messages. They are gratified that their hard work will be seen by everyone and are excited to work with the government."

City Campaigns

The importance of local and city campaigns in combatting climate change is clear, as evidenced by 350.org's Fossil Free campaign, which boldly claims: "City by city, town by town, we're ending the age of fossil fuels and building a world of community-led renewable energy for all." But how can cities play a role in replacing meat, dairy, and eggs with sustainable, plant-based options? There are three main policy areas in which city governments may implement plant-based climate solutions.

First, cities can capitalize on their procurement power and redirect funding towards plant-based food. Depending on the jurisdiction, cities can have discretion over spending in schools, senior centers, hospitals, and jails, and at city events and meetings. In 2021, the City Council of Berkeley passed a resolution to cut the amount of animal products purchased by the city by 50 percent by 2024, with the long-term goal of phasing out all purchases of animal products and replacing them with plant-based foods. Animal Save Movement joined a coalition of animal rights groups, led by DxE, advocating the resolution. Berkeley mayor Jesse Arreguín said, "This is a very important step for the city to take as part of our broader climate efforts, as well as building on our long tradition of promoting the humane treatment of animals here in the city of Berkeley."[59] In 2022, New York City mayor Eric Adams, dubbed the "Vegan Mayor," implemented Plant Powered Fridays in public school cafeterias. In addition, through Executive Order 8, he is revising standards for all food purchased and served by the city, bringing plant-based meals to city-run institutions such as jails and shelters, and offering financial incentives to neighborhood grocery stores to stock their shelves with healthy, plant-based foods.

Second, cities can promote community education on climate-friendly food and the co-benefits it produces for the environment, public health, and animals. Many cities already often encourage citizens to reduce fossil fuel use by cycling and using public transit, so these cities can easily add public information campaigns about the need for and benefits of plant-based food as well. For example, in New York City, Mayor Adams's Executive Order 9 dictates that "any advertising produced, published, or distributed by a city agency or placed on city property, that includes any representation or description of food, must feature healthy, whole foods such as fruits, vegetables, nuts, or whole grains."[60]

59 Anna Starostinetskaya, "Berkeley Becomes First US City to Commit to Vegan Meals," *VegNews*, July 21, 2021.

60 "NYC Mayor Adams Signs Two Executive Orders Promoting Healthy Food," Hunter College New York City Food Policy Center, February 22, 2022.

Third, cities can ensure that their investments promote plant-based foods rather than meat, dairy, and eggs. In 2013, Berkeley became the first US city to divest from the fossil fuel industry by joining 350.org's Fossil Free campaign. Eight years later, Berkeley passed a resolution urging its public employees' retirement system, which had $679 million in investments in twenty-eight operations in industrial animal agriculture, including companies like JBS and Tyson, to divest from the industry. As Nilang Gor wrote in an opinion piece: "Berkeley adopted a resolution to integrate environmental, social, and governance principles into the city's investing policy."[61]

The Plant Based Treaty calls for subsidies to be redirected towards plant-based food as a solution to the climate crisis. According to the Working Group II contribution to the IPCC Sixth Assessment Report (2022): "Subsidies directed at staple foods and animal-sourced foods could be shifted towards diversified production of plant-based foods in order to change the relative price of foods and thus dietary choice."[62] To target subsidies, we use digital media to make powerful actions easy, quick, and fun for individuals to participate in. In 2021, 1,200 people emailed President Biden to ask him to redirect a one-billion–dollar animal agriculture subsidy towards plant-based food solutions to the climate crisis. Always trying to keep up with the times, we created a "tweet storm" targeting the Biden Administration based on a hypothetical scenario involving a "meat comet," a parody concept we adapted from the Netflix film *Don't Look Up*.

Meanwhile, our chapters in Mexico are working with students and schools to garner support for the Plant Based Treaty, add plant-based meals to school menus, and update school curricula on climate change to include the role of dietary shifts . Our Plant Based Treaty teams constantly give talks, organize information tables with food sampling, hold film screenings, and put up posters and stickers. Our aim is to build a movement of Plant Based Treaty teams in schools across the globe,

61 Nilang Gor, "Opinion: Berkeley Urges CalPERS to Divest from Industrial Animal Protein, Factory Farming Companies," *Berkeleyside*, April 30, 2021.

62 IPCC AR6 WG2, *Climate Change 2022: Impacts, Adaptation and Vulnerability*, Chapter 5, February 2022.

with students and student unions leading the way in pushing for plant-based menus at school cafeterias and promoting only climate-friendly food at student-organized events. Further steps for students to explore include researching and launching divestment campaigns, following the example of 350.org, aimed at redirecting institutional and pension fund investments away from animal agriculture and towards plant-based food.

In yet other campaigns aimed at public education, our Plant Based Treaty posters and stickers have gone up in the streets of Mexico, Argentina, Uruguay, India, Italy, Israel, the US, the UK, and Canada. In February 2022, across Argentina, we rolled out a huge two-week vegan billboard campaign that was meant to parody the film *Don't Look Up*, promoting plant-based diets as a way to avert catastrophe as a giant meat comet headed towards Earth! We installed 760 billboards across five Argentinian cities—Buenos Aires, Mendoza, Mar del Plata, Cordoba, and Rosario. In addition, 10,000 posters and 16,000 stickers went up across cities all over Argentina.

The 3Rs

The third "R" (*Restore*) aims to actively reverse the damage climate change has done to the planetary functions, as well as Earth's ecosystem services and biodiversity. We are developing community gardens and coordinating seed and plant-based food giveaways. Our food giveaways help people try new and exciting, healthy, plant-based meals and encourage a dietary shift, while offering us an opportunity to do vegan outreach and collect Treaty endorsements. Our vegan outreach teams in the Netherlands regularly partner with Oatly to give away samples of oat milk to members of the public. In Buenos Aires, our team distributes over 500 vegan meals every week, along with educational materials about the benefits of a plant-based diet. In Montevideo, 100 homemade burgers—with everything, including the bread, patties, and carrot mayo, made from scratch—are distributed each week. In partnership with Million Dollar Vegan and Food Healers, thousands of meals have been distributed in Mumbai. There are similar initiatives in multiple cities, including Ankara, LA, Guadalajara, Mexico City, Toronto, and Tel Aviv.

We are working on community gardens, school gardens, and home gardens in backyards, on balconies, and indoors. Community gardens can be placed next to food banks and primary healthcare facilities, in parks and animal sanctuaries. In 2021, in Buenos Aires, two community gardens were installed, the first of many planned across the city as part of a Plant Based Treaty project. Also as part of our program, local communities are taught how to grow and care for vegetables, as well as how to cook delicious vegan dishes.

We have set Treaty endorsement targets for 2023 to coincide with the very first global stocktake of the Paris Agreement. Therein lies an opportunity for the Plant Based Treaty. If we build enough momentum and create enough bottom-up pressure from the public, scientists, businesses, and municipal governments, we can shape policies and help set the pace of reform, first by persuading national governments to negotiate a global plant-based treaty. By the end of 2023, we will have new endorsement targets for the period of 2024 to 2029, to be reviewed again after the next global stocktake occurs.

We saw the need to get vegan food onto the menu in our communities and at the Conference of the Parties, and we just went for it. At the Bonn Climate Conference, we got our initiative to be acknowledged by working outside the box, by telling it like it is in our position paper and simply providing delegates with vegan food instead of waiting for the conference building's cafeterias to offer climate-friendly vegan menus. We are part of the conversation now, more than ever before, when our activism focused only on animal vigils. We have been bearing witness for over a decade, yet have had few conversations with policymakers. That has all changed. Now, we need to figure out how to show the world that whether we can solve the climate crisis depends on how we see and treat farmed animals, and all other creatures who call Earth home. City to city, town to town, we need to boldly proclaim our message, like the enlightened king in Tolstoy's moral tale "Esarhaddon, King of Assyria": "Afterwards, he went about as a wanderer through the towns and villages, preaching to the people that all life is one, and that when men wish to harm others, they really do evil to themselves."

10

Final Words

RICHARD HOYLE'S PIONEERING WORK AT The Pig Preserve provides a blueprint on how to provide a natural home for liberated farmed animals. We can—and must—find a way to share the Earth with the other creatures who call it home.

I didn't know a lot about pigs until I visited The Pig Preserve. I had known mostly pigs in transport trucks—millions of them since we started holding weekly vigils. To me, they are very similar to my dog, Mr. Bean, who was my best friend—highly expressive and attentive, with big personalities.

We all do share this moral responsibility "to not look away but to come closer," to try to help any suffering creature. In the words of my animal activist friend Joanne O'Keefe: "Look into their eyes, together with us, and strengthen your dedication to fighting for them with all your heart. Look into their eyes, and know you're doing the right thing by exposing the truth to others. Look into their eyes, and feel the hell they know as life. Join us, bear witness, and tell the world what it refuses to see."

We need everyone to join animal vigils and to visit sanctuaries. If everyone came to bear witness to the suffering of pigs at slaughterhouses, we could change the current dominant worldview. If everyone came to visit sanctuaries like The Pig Preserve and got to know pigs on their terms, we would respect them as fellow travelers on our precious planet, and we would not want to eat them—or any other sentient being.

Mr. Bean, a co-founder of Toronto Pig Save, stands near the entrance of Quality Meat Packers slaughterhouse as an empty transport truck leaves the site. We started holding three pig vigils a week in 2011 shortly after Mr. Bean was adopted. (photo: Anita Krajnc)

Sad pig in transport truck taken at a Toronto Pig Save vigil. (photo: Anita Krajnc)

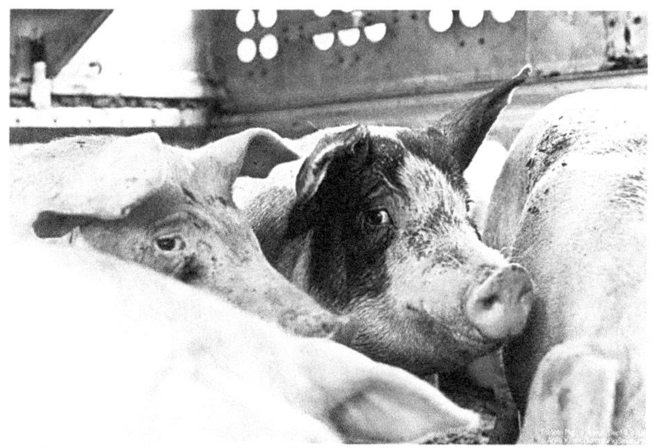

Piebald pig amongst a group of despondent pigs in a transport truck witnessed at a Toronto Pig Save vigil. (photo: Anita Krajnc)

A group of Toronto Pig Save activists, including Caroline Wong (forefront) who joined our "Big Pig Trip" to the Pig Preserve, bear witness at a pig vigil on "Pig Island" at a busy intersection in downtown Toronto. (photo: Anita Krajnc)

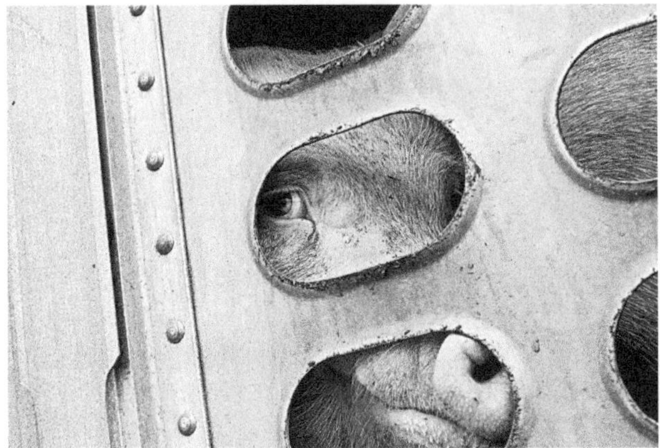

Utterly distraught pig looks out at us from a moving transport truck as lights turn green on "Pig Island." (photo: Anita Krajnc)

A pig climbs on top of another because there is nowhere to go in the crowded and hot transport truck heading to Fearmans slaughterhouse in Burlington, Ontario. (photo: Anita Krajnc)

Toronto Cow Save vigil participants bear witness to and document frightened and exhausted cows in a transport truck outside of a Toronto slaughterhouse. (photo: Louise Joregensen)

Arriving at Farmer John slaughterhouse near downtown LA, after traveling through the extreme heat of the CA desert in the back of a truck, this pig was foaming at the mouth and trying to get a drink of water from the LA Animal Save activists. (photo: Bobby Sudd)

A woman pets a piglet at a Medellin Animal Save vigil in Colombia in 2018. (photo: Azul Cardozo)

Activists link arms to form a hug around a truck and transmit their love to the pigs at a Medellin Animal Save Vigil in 2017. Azul says, "Their pain is our pain. We will continue to be witnesses so that the reality that they live reaches every person." (photo: Azul Cardozo)

Giving water to thirsty and exhausted piglets in the oppressive heat of Río Negro in Colombia. (photo: Azul Cardozo)

Regan Russell standing with her homemade sign at the front gates of Fearmans slaughterhouse meters away from where she was struck and killed by a pig transport truck on June 20, 2020. (photo: Agnes Cseke)

Plant Based Treaty table at the first Burlington VegFest.
(photo: Anita Krajnc)

Aprajita with city of Jamnagar Mayor Binaben A. Kothari after she endorsed the Plant Based Treaty. (photo: Naman Dedhia)

Photo of Anita giving a presentation on the Plant Based Treaty at the Burlington VegFest in July 2022 (photo: Agnes Cseke)

Afterword

pattrice jones

TRUFFLES THE PIG CAME TO VINE Sanctuary because she loved birds. When I met her, she was already an elder. She had spent her life at an informal sanctuary that had lost access to its land. We had already agreed to take the chickens and guinea fowl from that refuge when we learned that there was also a pig who had never known other pigs and who was deeply attached to her feathered friends. Rather than ask her to go alone to someplace like Pig Preserve, where she might be stressed by navigating new social dynamics while mourning the loss of everyone she had ever known, we agreed to figure out a way to accommodate Truffles too. We scrambled to use whatever we had at hand to build Truffles her own homestead—which she immediately turned into a kind of refuge within a refuge, inviting birds who wanted a respite from the hustle and bustle of flock life to nestle within the elaborate straw structures she rebuilt every day.

That spirit of improvisation in solidarity with kindred of other species infuses the narratives of both Animal Save Movement founder Anita Krajnc and Pig Preserve founder Richard Hoyle. Both were so moved by encounters with pigs that they upended everything about their lives. Each decided, at various junctures, to try to do something that hadn't been done before, despite having no way to know in advance how it would go. They are very different people who elected to jointly publish a book from which we all can learn; from whatever standpoints or life experiences we have come to their stories.

The first thing we can learn is that, like Truffles, every pig is "one of a kind." The next thing we learn, if we didn't already know it, is that pigs are treated like disposable objects, most egregiously by the pork industry but also by purveyors and purchasers of pigs as pets. As an antidote, this book offers us numerous inspiring and instructive examples of people using their own unique skills, resources, and insights to care for pigs and push back against those who harm them for pleasure or profit.

Truffles came to VINE because she loved birds, but the only reason she had to go anywhere was because yet another small sanctuary had fallen. Richard writes of the struggles that he and Laura encountered in maintaining the project they founded. Such struggles are always both material and emotional. Just as activists put their emotional and physical well-being on the line at vigils, sanctuary operators must cope with ceaseless physical and emotional labor, with no time to properly grieve inevitable deaths. The persistent problem of where to find funds adds anxiety to anguish.

Many smaller sanctuaries are in need of your support right now. Just today, while taking a break from writing this piece, I came upon an urgent call for help from a sanctuary founder who has run out of energy and funds and will need to begin the process of trying to place animals at other sanctuaries—all of which are always full. I literally cannot count the number of times that we have taken or helped to place animals from sanctuaries that ran out of money.

The situation has become even more grim in recent years, due to the efforts of adherents of Effective Altruism to convince donors and grant-makers that support for animal rescue and care squanders funds that should be spent instead on promoting veganism and bringing new meat analogs to market. Grant opportunities decreased just as the number of new sanctuaries increased, putting more fiscal stress on everyone.

You can help. Follow sanctuaries on social media and share their fundraising posts. If you find that seeing photos and videos from a particular sanctuary enriches your life and boosts your own well-being, pitch in with a donation or an offer to volunteer. Setting up a monthly donation, even in a small amount, can really help. If a sanctuary can count on you to buy just one gallon of tractor fuel, bale of straw, or sack of feed each month, that's one less thing they have to worry about.

Don't hesitate to reach out with emotional support too! If you see on social media or in a newsletter that a sanctuary resident has just died, send your condolences. If you notice that a sanctuary you follow is in the pathway of a storm, let them know that you're thinking of them and check in afterwards to make sure they're OK. There are many more resources available for sanctuaries than there were when Richard and Laura (and I) were starting out, but founders of new sanctuaries don't always know about them. If you see a small sanctuary struggling, tell them about the Open Sanctuary Project and the Global Network of Farm Sanctuaries.

I don't imagine that Richard and I agree about many things. He's politically a conservative who is proud of his past military service. I'm a lesbian ecofeminist who sees patriotism and patriarchy as two sides of the same coin, with that coin being the profits of violence and exploitation. And yet, here I am writing the afterword to a book he coauthored and which I am glad you have read. That's because everything depends upon finding ways to work together on the things about which we agree, even when we have substantial disagreements.

Empathy helps that happen. I can relate to so much of what Richard has written. He includes an hour-by-hour listing of a day in 2000, during which he spent most of the day caring for rescued pigs. I have my own hour-by-hour listing of a different day in 2000, during which I spent most of the day caring for rescued chickens. Like Richard, I was constantly tripping over cats—if you run a sanctuary in a rural region, then cats will magically manifest, regardless of whether or not cats are within your intended remit—and I know all too well how it feels when someone shows up at your doorstep with an animal they tell you they will kill if you do not agree to take them immediately. Like Richard, I have scrambled to cope with incessant demands to take in animals surrendered by humans who obtained them due to fads—for Richard, pigs discarded by people who bought them as pets; for me, unwanted roosters discarded by backyard hen-keepers who buy chicks from hatcheries. In both cases, the numbers of animals paying the price for human follies is almost unimaginable.

I laughed out loud at Richard's "you might be a sanctuary director if. . . " list. All of my clothes also bear mysterious tears and stains. He's probably familiar with one of my favorite guessing games, "how did I get that bruise?" Richard tells us about another question—"Are they happy?"—that led him to make a big change and which also has kept me awake at night.

Different standpoints lead to different insights, so let me share something I noticed in Richard's narrative. He tells us of the shockingly high proportion of surrendered pigs who needed refuge due to domestic violence. He recounts a story (all too familiar to me) of a woman pleading for him to take a pig who her husband has threatened to kill and eat. And he tells us that female firefighters were much more open to his vegan meals than his more macho comrades. All of these suggest to me that we cannot avoid thinking about gender when wondering how to help pigs.

Meat-eating is one way that men perform their masculinity. Demonstrating dominance over animals by means of hunting, fishing, bull-fighting, cockfighting, dog fighting, or riding in rodeos is another way that men prove their masculinity to each other. Boys and men who refuse to participate in such activities may be called "sissy" or a homophobic slur. Domestic violence, whether against partners, children, or companion animals, is another way that men who have grown up believing that masculinity means dominance demonstrate their "manhood."

All of these well-known and easily observed conjunctions make it essential for men involved in animal rescue or advocacy to be aware of gender and to interrogate their own ideas about masculinity. Otherwise, it can be easy to fall into the trap of seeing yourself as a hero leading the charge for animal liberation, rather than seeing yourself as the ally of animals who are always actively seeking their own freedom. In sanctuary work, it can be particularly important to demote yourself from hero to helper, so as to become better able to foster the self-determination and freedom that Richard so ardently sought to enable at Pig Preserve. It's also important to be aware of the degree to which our own gender stereotypes and presumptions of heteronormativity can influence how we see animals. I wonder if knowing that pigs are among the many animals whom same-sex relationships are not uncommon, might have caused Richard to reconsider seeing every male pig as "genetically programmed" to pursue "mating" and whether that insight might have allowed him to develop even deeper relationships with them.

Relationships are the key to both sanctuary work and activism. I imagine that many people may be most impressed by Anita Krajnc's individual heroism in going on trial for giving water to a pig or her individual creativity in seeing the possibilities of staging vigils outside of that particular slaughterhouse, but what I notice in Anita's narrative is her genius for drawing upon and building relationships. She reads widely, making connections across the books she has consulted. She consults people in conversation too, and encourages her comrades to discuss among themselves. I can testify to this not only as a reader of this volume but also as someone who has had hours-long discussions with Anita at her initiation, swapping stories, brainstorming, sharing book recommendations, and *thinking together* about the problems at hand.

I'm so grateful to Anita for making her way of thinking more accessible to other activists by co-authoring this book. Her way of working is instructive, and not only for those of us who are actively working against animal exploitation and for climate justice! If you know someone who is engaged in any kind of activism, do them the favor of recommending that they read her portion of this book. If you're an activist of any kind, encourage your comrades to read it, and then get together to discuss how you might put its lessons into action.

In addition to the focus on relationships, two other elements of Anita's narrative stand out for me. First, I cannot emphasize strongly enough how important in-person, local community organizing always has been and always will remain, *especially* in the context of a world that has become ever-more warped by the distractions and polarizations enabled by social media. This is not to say that online activism has no place. As Anita notes, it can be vitally important. But there is no substitute for talking with people in person, not only in order to more effectively influence the people whose behavior you hope to change but also in order to gather more people for the cause. As Angela Davis often notes, true social change always comes from the bottom up and requires regular people to become collectively aware that they themselves have the power to change the cultures in which they participate. A society is an aggregation of people. As people change, one by one by one, societies can shift.

Organizing is a skill like any other. You've already learned a lot by reading this book, and you can learn more by checking out some of the books Anita mentions, especially those by Saul Alinsky. Alinsky was known not only for effective organizing but also for creative tactics, which brings me to the second element of Anita's narrative to which I want to draw your attention: improvisation. Anita quotes Sue Coe, who says "You don't need money, you don't need credentials. Just placing yourself in the vortex creates change. That's the beginning of change."

Truer words were never spoken. Every slaughterhouse sits in a specific place as a result of a confluence of material, economic, and social factors, all of which are always in flux. Every purchase of a pork chop is similarly situated, with a variety of factors converging to lead the purchaser to make that choice that day. This is true of every problem you might want to solve. Adding yourself to the mix changes the mix.

You don't have to have everything all sorted out to begin, but it's a good idea to take Anita as a role model for how to constantly ask what effect your efforts are having, making adjustments along the way. New ideas can come from anyone, so be sure to engage in collective imagining. Because people are more than just brains, go beyond verbal messaging to include art, movement, and other ways to spark people to really *see* and have empathy for nonhuman animals.

Whichever tactics you choose, notice how helpful it was for Anita's team to visit sanctuaries. That's an excellent example of how different forms of activism can complement one another. To create the kind of substantial, sustainable change that animals and the planet need, we will need to approach problems like slaughterhouses from a variety of angles using a variety of tactics. If we can do that in concert rather than in competition with each other, we will be much more likely to succeed.

People who are anguished by the exploitation of animals, climate change, or any other significant social problem often wrestle with the question, "But what can *I* do?" The answer is always that the problem is multifaceted and there are more things to do than people to do them, so the best thing for *you* to do is whatever your own unique constellation of skills, resources, and relationships puts you in a good position to do. That might mean offering your time to an existing project such as the Plant-Based Treaty or it might mean starting something new, as both Richard and Anita have done. Either way, approach the work with the same spirit of informed improvisation that runs through this book. Educate yourself, but don't wait until you know everything you need to know (spoiler alert: that day never comes) before taking action. Understand that you can't possibly know in advance what will happen, and be prepared to "try, try again" in different ways until you hit upon something that works. Even if something seems to be working, be aware that circumstances are always changing and constantly reassess, as both Anita and Richard did throughout.

Most importantly, always remember the pigs (and trees and frogs and people who are struggling to help others even as they cope with their own difficulties), who maintain a remarkable capacity for generosity and growth despite everything they have undergone. The impulse that led Richard and Laura to start a sanctuary, Anita to found a worldwide movement, and Truffles to offer her own home as a refuge for birds is rooted in a deep desire for connection shared by all social animals. That wish for more healthy relationships—with nature, with other animals, and with other people—lives in you too. . . and in the heart of every person whose behavior you hope to change. Draw upon that desire within yourself and call to it in other people as you do whatever reading this book inspires you to do. I don't presume to be able to speak for animals, but I feel quite certain that's what Truffles would want you to do. In her memory, and in remembrance of the courage and kindness of Anita's comrade Regan Russell, I thank Anita and Richard for sharing their stories and thank you in advance for whatever you will do in response.

pattrice jones is a cofounder of VINE Sanctuary, an LGBTQ-led refuge that works for social and environmental justice as well as animal liberation. Hundreds of formerly farmed animals co-create a multispecies community at the sanctuary, which was the first to rehabilitate roosters used in cockfighting and which is the subject of *The Oxen at the Intersection*, published by Lantern in 2014.

Compassion is Not a Crime

Final Words from Moby

I GREW UP IN THE 70s eating the standard American diet. Lots of Burger King, meatloaf, salami sandwiches, etc. Then in 1984, I realized that it was incompatible to love animals and also contribute to their suffering and death. I had rescued a cat named Tucker from the dump in Darien, CT and he taught me that all beings are individuals with their own lives to lead; that everyone wants to be free from suffering and pain and live a free life. So, I became a vegetarian in 1984, and then a vegan in 1987.

Becoming vegan was very transformational for me, it allowed me to live a life with an ethical purpose, a life of cohesion and consistency. It has kept me healthy despite struggles with alcohol and it has given me clarity and purpose.

I believe if humanity were put on trial, our greatest crime would be responding to innocence and vulnerability with violence and cruelty. When I heard about Anita Krajnc being charged for giving water to thirsty pigs enroute to a slaughterhouse on a scorching hot day, I found it incredulous: the opposite of what should have happened was happening. Compassion is not a crime, yet Krajnc, who should have been applauded for her act of compassion, was arrested and charged, and initially at least faced up to 10 years in prison. This inspired me to attend a vigil here in Los Angeles and it brought me back to all those years ago meeting and rescuing Tucker the cat. Meeting the victims of animal agriculture in the trucks, those innocent defenseless pigs being needlessly exploited and killed, is a truly transformational experience. It is no wonder to me that attending animal vigils has spread all around the world.

What also struck me bearing witness to pigs on transport trucks is how different they are to pigs on sanctuaries. I've been fortunate enough to spend many days visiting sanctuaries and meeting animals in a relaxed and tranquil environment. They look so different to those you meet on the trucks. There is no terror and anxiety, no suffering. Their faces and eyes are different. They are content, at ease. I only wish people who consume animal products could see that difference, could witness the cruelty they are contributing to.

First and foremost, my veganism is based on my belief that all animals are entitled to their own life and their own will, and that I can't in good conscience be involved in contributing to, or causing, an animal's suffering. However, the fact that animal agriculture is a leading cause of rainforest deforestation and climate change is another undeniable reason why we must stop killing animals and go vegan. Despite a lot of animal charities beginning to address this, I find it strange how there is no charity specifically working on animal agriculture and climate. So, I was delighted to learn about the Plant Based Treaty and was happy to endorse it. While fossil fuels are clearly a huge part of the problem, we simply cannot solve the climate crisis without going plant-based. Change is imperative, and I endorsed the Plant Based Treaty for the animals, the planet, and human health.

Even if animal agriculture wasn't so destructive to the environment and didn't involve killing animals, it would still make no sense as the future of food as it makes people fat and sick and is a horrible use of resources. We should also remember that the consumption of animal products is the leading cause of antibiotic resistance, and a leading cause of diabetes, obesity, heart disease, and many cancers.

I've been a vegan for 35 years, and I honestly never thought that veganism would be as vibrant and widespread as it is today. When I became vegan in 1987 there were maybe 5 vegan restaurants in the whole world. Now there are at least 10 vegan restaurants within a 5-mile radius of my house in LA. Supermarkets are full of vegan food options, it really has hit the mainstream in a big way over the last decade. It is easier than ever to be vegan.

We can still do more though. Governments need to stop subsidizing the animal agriculture industry, which is one of the demands that the Plant Based Treaty calls for. I'd like to see this applied to all industries that are causing damage to our society: food, energy, guns, pharmaceuticals, chemicals, etc. It's patently absurd that trillions of dollars in tax revenues go to industries that make us sick or kill us.

I personally believe that humans are a broken species. No other animal is so separated from the divine or whatever life source you want to name. No other animal exploits and kills for fun or has devised a system as cruel as factory farming, so completely removed from nature. I see all other animals and creatures, even the violent ones, as being in a state of connection with God, and when we inflict violence and torture upon them, we are acting out as creatures separated from the divine. When you see an animal being abused or tortured or killed by a human, when you meet them on transport trucks bound for the slaughterhouse, you can see, in their eyes, not just terror, but utter bewilderment. The time is long past for us to stop, take a step back, and move forward with compassion for animals, our health, and the planet.

Moby is an artist, musician, singer, songwriter, videographer, producer, vegan restaurateur, author of four books, animal advocate, and humanitarian. He has sold over twenty million records worldwide and is credited with helping to bring dance music to a mainstream audience. Moby is a staunch animal rights activist and long-time vegan supporting numerous causes through his work and advocacy.

About the Authors

Richard Hoyle received his BA/MA from the Citadel and The Naval Postgraduate School in management and education. After serving our country as a Marine, Rich became a career firefighter (promoted to captain and chief), and a paramedic. He founded the Pig Preserve in 2006 with his wife Laura, and has provided loving care, community, and sanctuary to thousands of pigs consistent with their natural environment.

Anita Krajnc is the co-founder of Toronto Pig Save and the Animal Save Movement, a worldwide network of Save groups bearing witness to farmed animals and promoting veganism and love-based, grassroots activism. Anita received her Ph.D. in Political Science from the University of Toronto and is a staunch follower of Leo Tolstoy and Mahatma Gandhi. Anita has also been an assistant professor at Queen's University (Ontario) and resides in the Toronto, Canada area.

About the Publisher

Lantern Publishing & Media was founded in 2020 to follow and expand on the legacy of Lantern Books—a publishing company started in 1999 on the principles of living with a greater depth and commitment to the preservation of the natural world. Like its predecessor, Lantern Publishing & Media produces books on animal advocacy, veganism, religion, social justice, humane education, psychology, family therapy, and recovery. Lantern is dedicated to printing in the United States on recycled paper and saving resources in our day-to-day operations. Our titles are also available as ebooks and audiobooks.

To catch up on Lantern's publishing program, visit us at www.lanternpm.org.

facebook.com/lanternpm
instagram.com/lanternpm
twitter.com/lanternpm